New Shock-Resistant Materials for Work Equipment used in Potentially Explosive Atmospheres

Romeo-Gabriel CHELARIU[1]

Ramona CIMPOEŞU[1]

Gabriel-Dragoş VASILESCU[2]

Costică BEJINARIU[1,3]

[1]Faculty of Materials Science and Engineering, "Gheorghe Asachi" Technical University of Iasi, 67 Dimitrie Mangeron Street, 700050 Iasi, Romania

[2]National Institute for Research and Development in Mine Safety and Protection to Explosion—INSEMEX, 332047 Petrosani, Romania

[3]Academy of Romanian Scientists, Ilfov 3, 050044 Bucharest, Romania

Published by **Materials Research Forum LLC**
Millersville, PA 17551, USA

Published as part of the book series
Materials Research Foundations
Volume 186 (2026)
ISSN 2471-8890 (Print)
ISSN 2471-8904 (Online)

Print ISBN 978-1-64490-386-5
ePDF ISBN 978-1-64490-387-2

Distributed worldwide by

Materials Research Forum LLC
105 Springdale Lane
Millersville, PA 17551
USA
https://mrforum.com

Printed in the United States of America
10 9 8 7 6 5 4 3 2 1

Table of Contents

Materials Research Foundations **186** (2026) https://doi.org/10.21741/9781644903872

CHAPTER 1

Current State of Research on Non-Sparking Alloys with Applications in Potentially Explosive Environments

Romeo-Gabriel CHELARIU[1], Ramona Cimpoeşu[1]*,Gabriel-Dragoş VASILESCU[2],Costică Bejinariu[1,3]

[1]Faculty of Materials Science and Engineering, "Gheorghe Asachi" Technical University of Iasi, 67 Dimitrie Mangeron Street, 700050 Iasi, Romania

[2]National Institute for Research and Development in Mine Safety and Protection to Explosion— INSEMEX, 332047 Petrosani, Romania

[3]Academy of Romanian Scientists, Ilfov 3, 050044 Bucharest, Romania

ramona.cimpoesu@academic.tuiasi.ro

Abstract

The need to reduce the number of explosions and flash fires in the workplace is driven by humanitarian and economic considerations. The humanitarian concerns are obvious: explosions and fires can cause extremely serious injuries and fatalities. Eliminating the sources of sparks (electrical or mechanical) that cause explosions can significantly reduce the risk of deflagration, particularly in environments that are prone to them.

Keywords

Non-Sparking, Current State, Cu Alloys, Explosive, Risks

1.1 Non-sparking materials

Electrical and mechanical sparks are the main source of ignition in various situations in machinery and production warehouses [1]. In order to minimise the intensity of sparking, systems are usually manufactured from non-sparking metallic materials or non-metallic materials if tolerated by the specification. A particular feature of all non-ferrous alloys used in potentially explosive environments is that they have high thermal conductivity. Cu-Al materials, stainless steels, Cu-, Ag-, Al-based alloys and galvanised steel are among the few materials that have shown no sparking or cold sparking. As well as being less prone to sparking, these alloys are also more resistant to corrosion, allowing longer service in a variety of gaseous or liquid environments. Due to their lower strength compared to steels, non-sparking inspections should be periodically performed to determine the level of material degradation. These tests can now be easily conducted using surveillance cameras with specialised detection software. For various applications where specific strength is not required, non-metallic materials are usually used for systems operating in potentially explosive environments [2]. These materials are based on plastics, wood, thermoplastic polymers, leather, but also special metallic materials in general, currently based on Cu-Be, etc. However, whilst the characteristics and properties that lead to spark formation and growth are not really known, there are few metallic materials that are considered to be spark free. These materials are based on binary Cu-Zn, Cu-Sn, Cu-Ni, Cu-Al, Cu-Be and Ti-based alloys, as well as ternary systems of their combinations [3]. These materials have lower mechanical properties than Fe-C materials, but can replace them in many applications. Naturally, the strength of these materials is lower, but with proper geometric design and machining processes, they can successfully replace steels or cast irons. Al-based materials are also covered by this standard, but it has not yet been established that aluminium is a non-sparking material [4]. For a material to be considered non-sparking it is not sufficient to use a soft alloy. If it is considered that the sparking process is related to the removal of material parts during mechanical contact, then the ductility of the material should be considered more than its mechanical strength. Aluminium is still a qualified candidate for listing as a non-sparking material [5].

Nickel and its alloys produce a wavy orange streak of low-volume sparks during the spark test and it is claimed that it could become a non-sparking material, but these properties are not fully substantiated. In a spark generation process, the ignition of gas, aviation gas supply, paraffin and pre-heated oil by spark ignition was analysed by burning various materials over the 'rock'. The materials used were Al, Ti Mg, Cr-Mo and stainless steel. All Al-based alloys can create an explosive environment. Experimental tests have been proposed to ignite a 7% CH_4-air combination to analyse the heat on the top surface of the starting material to produce sparks. An area of 3x2 mm x mm is required to ignite a gaseous combination at a temperature of 1300°C, while the area required is reduced to 2x1.5 mm square at a temperature of 1550°C. Lowering the source temperature below 1100°C requires an area of 2x10 mm^2 or more [6].

For brass and bronze (Cu-based alloys), their non-sparking properties are important, but another important property is their impressive thermal conductivity. Cu and Cu-based alloys have a good chemical composition and structure distribution and are often single-phase materials. Normally these materials are not worked by heat treatment and the main properties can be modified by cold or hot plastic deformation. However, when Be (beryllium) is used as an alloying element between 0.6 and 3 %, Cu alloys exhibit hardening by precipitation. Cu-Be alloys are usually containing Be up to a limit of 3-4%, but their processing by various methods can introduce contaminating elements into the working atmosphere and become harmful to workers in the field. OHSA - The Occupational Health and Safety Administration has set a limit of 1.95 g/m3 of Be in the warehouse working atmosphere over a 7-10 hour working period, as Be is considered hazardous to humans [7]. Cu alloys with high levels of Be can cause illness, and prolonged exposure to such environments can cause headaches and dizziness.

Explosion protection is particularly important for safety as explosions also endanger the health of workers due to the uncontrolled effects of flame and pressure, harmful reaction products and the air inhaled by workers. The CuAlBe alloy is proposed as a solution for mechanical actuators, such as gears, operating in potentially explosive atmospheres.

Produced from a CuBe base alloy and pure aluminium in an induction furnace, the material has large grains in the cast state. After hot rolling (600 seconds at 900 °C), a slight variation in chemical composition due to oxidation of the material, small cracks at the edges and a preferential orientation of the grains along the rolling direction were observed [8].

Copper, one of the most important metals used in industry, is widely used both in its pure form and as an alloy. Its value has increased mainly due to the special physico-chemical properties of its various combinations with other metals, but also because it is irreplaceable in some practical applications. Aluminium bronzes are the most common and valuable type of special bronze because of their superior properties. Aluminium bronzes can be binary (simple) alloys, where copper is alloyed only with aluminium, or complex bronzes, where they contain other alloying elements in addition to aluminium, such as iron, manganese or nickel.

Beryllium bronze belongs to the group of so-called dispersion hardenable alloys. They are characterised by the temperature dependence of the alloying element solubility. One of the applications of beryllium (Cu-Be) bronzes is as a non-sparking material for applications in potentially explosive environments (natural gas or mining) [9]. When quenching is performed from a single-phase section, an excessive number of atoms of the main alloying component is formed compared to the equilibrium state of such a system. The resulting concentrated solid solution is characterised by thermodynamic instability and a tendency to decompose; this process is activated as the temperature level increases. The densification effect is explained by the dispersion of the precipitates obtained as a result of the decomposition of the substances.

All bronze alloys that contain beryllium in their structure are characterised by high temperature resistance - their products can operate at temperatures of up to 340°C without altering their mechanical properties, and when heated to 500°C, the mechanical properties of all beryllium bronzes become equal in performance to both aluminium and tin-phosphorite compositions at a standard operating temperature of around +20°C. This property allows beryllium bronze to produce castings of the highest quality. Beryllium alloys are easily adaptable to any mechanical processing (cutting, brazing and welding). For example, all beryllium alloys must be brazed immediately after mechanical pickling. In this case, it is essential to use a silver alloy and a welding flux. It is important that fluoride salts are always present in the flux itself. Over the past few years, vacuum brazing has become very popular - brazing is performed under a thick layer of flux. This ensures a unique product quality. More recently, innovative processing routes have been proposed to produce different types of Cu-based alloys with improved functionality and/or mechanical properties [10]. CuAlBe alloys exhibit a solid state transformation from an austenitic to a martensitic structure on cooling and vice versa on heating [11].

1.2 Hazardous explosive environments in industry

Explosion protection is especially important because explosions endanger the life and health of workers due to the uncontrolled effects of flame and pressure, harmful reaction products and the air breathed by workers. An explosion occurs when a fuel mixed with air (i.e. a sufficient amount of oxygen) reaches the explosive limit in the presence of an ignition source. If a hazardous explosive environment can form, explosion protection measures are required. The first step is to avoid the formation of potentially explosive atmospheres. Table 1.1 below gives some examples of potentially explosive situations in various industrial environments [12].

If the formation of hazardous explosive atmospheres can be prevented, no further action may be necessary. In many cases it is not possible to avoid explosive atmospheres or ignition sources to a sufficiently safe level. In such cases, measures must be taken to limit the effects of explosions to an acceptable level. Explosions can occur in fires.

Table 1.1. Examples of potentially explosive situations in different industries.

Illustration	Industry	Risk examples	Reference
	Chemical industry	The chemical industry uses many techniques to convert and process gaseous, liquid and solid flammable substances. These processes can create explosive mixtures.	[13]
	Warehouses and civil engineering	Warehouses can generate flammable gases. Extensive engineering measures are required to prevent their uncontrolled release and ignition. Flammable gases from various sources can accumulate in poorly ventilated tunnels, basements, etc.	[14]
	Energy suppliers	Possible ignition dusts/gas mixtures which may occur as lumps in the transport industry, coal preparation or drying (these materials are not explosive in combination with atmospheric air).	[15]
	H_2O purge marks	A possible source of an ignition atmosphere is fermentation compounds released during wastewater treatment.	[16]
	Gaseous compounds distribution	Explosive gas/air mixtures can form in natural gas leaks.	[13]
	Wood processing	Woodworking produces wood dust, which can be explosive or form hazardous mixtures.	[17]
	Painting workshops	Excessive spray paint that builds up in spray booths as spray guns coat surfaces, as well as removed solvent vapours, can create explosive environments.	[18]

In some cases, explosions are followed by fires; in others, only the explosion occurs. In order to prevent this, in accordance with the regulations of art. 7(3) of Legislative Decree no. 319/2006 and ensuring protection against explosions, the employer must take technical and/or organisational measures according to the nature of the activity, in order of priority and in compliance with the following basic principles (319/2006):

a) analysis of potential ignition behaviour in explosive environments;

b) here it is not possible to prevent the formation of explosive atmospheres, it is necessary to use non-sparking materials to avoid the possibility of ignition.; and

c) limiting the harmful effects of an explosion to ensure the health and safety of workers.

1.3   Friction spark accidents

In many industrial incidents, the main problem has been the occurrence and development of friction sparks. In most cases, a complete analysis of the sparking process has not been undertaken. There are a few factors that need to be considered, such as the friction degree of two or more materials in contact, the intensity of the impact and the amount of shock, which can become important and influential for the atmosphere in which they work [19]. Gears, especially those with high loads, can be considered to have a high risk of sparking and spark propagation in the contact function. An industry with a high risk of explosion is natural gas production and fuel transport and processing. One example [20] mentions a fire during the handling of fuel preparation equipment. Sparks were caused by a heavy metal tool during manual work on a metal engine. The final report did not suggest the use of non-sparking tools. Observations were made on the tool and the main conclusions focused on equipment handling. A better solution may be to use a non-sparking material with similar mechanical properties [21].

Another example dates back to 1854, when a caravan of horses pulling carriages started to burn after coal dust was ignited by sparks from horseshoes in contact with the road. In 1979, a large gas pipe caught fire from friction sparks caused by manual cleaning with spark-producing metal tools.

Most incidents show that the condition of the material (dust, rust, etc.) and the contact between different materials under different stresses play an important role in sparking. Wear and friction of metallic materials can cause sparks that lead to fire or explosion ignition [22].

1.4   Cold sparks and best practices to avoid explosions

Contrary to popular belief, the non-sparking process does not completely eliminate sparking. In some cases, such equipment can produce 'cold sparks'. In this case, they are insufficient to start a fire due to the small amount of heat they generate. These sparks are considered safe for the working environment. For a spark to be hazardous, it must have a high temperature and a certain amount of time to start burning a flammable compound [23]. Following best practice for non-sparking, non-explosive equipment can ensure that its non-sparking properties are well maintained. A general conclusion is that such equipment will be placed in a different location from Fe-based materials. In addition, the elements should be kept away from other materials and certainly not in direct contact with acetylene, as there is a possibility of highly volatile and hazardous materials being formed. In any case, the venting process is very important to prevent the development of hazardous compounds from 'clumping' or building up around the sparking elements [24].

1.5 CuTi alloys as an alternative to CuBe alloys

CuTi alloys are increasingly being investigated as very high strength conductive materials for applications such as arc suppression, interconnects, spark arrestors, etc., essentially replacing the range of conventional CuBe alloys. This attempt to replace CuBe-based alloys was largely triggered by the recognition of the health hazards associated with the metallurgical production of Be-based alloys. CuBe alloys are considered as typical elastic alloys and are generally used for special applications due to their high performance properties [25,26]. However, the toxicity and high cost of Be significantly limit future applications [27]. In recent decades, many studies have been conducted to find a good substitute for CuBe alloys, and CuTi alloys have received much attention [28]. In addition, Cu-alloyed Ti exhibits some unique properties such as good strength, strong biocompatibility and acceptable corrosion resistance, and therefore Ti-Cu alloys have been identified as a new antibacterial biomaterial [29].

Age-hardening of copper-titanium (CuTi) alloys containing approximately 1-5 wt% Ti (1-6 wt% Ti) has been known since the 1930s. The mechanical and physical properties were found to be comparable to those of the widely used copper-beryllium (Cu-Be) alloys, with better resistance to high temperatures and superior external stress damping behaviour. However, the electrical conductivity of the alloys falls slightly below that of Cu-Be alloys after a period of use.

Many electrically conductive thin sheet springs are made from age-hardened copper-titanium alloys due to their excellent mechanical strength and electrical conductivity. Such arcing sheets are usually produced by a process involving the production of a copper-titanium melt, its casting, hot working of the cast CuTi alloy followed by alternate annealing and cold working to a final shape, where the alloy may be solution annealed or age hardened by cold working. Its structure, heat treated by solution hardening, results in an average grain size of at least 40 microns and even up to 100 microns.

The age-hardened copper-titanium alloy was developed as a low-cost replacement for the better-known copper-beryllium alloy, also age-hardened, disclosed in US Patent 4.425.168 granted to Goldstein et al. in 1984. However, the conventional age-hardened copper-titanium alloy is unsatisfactory and needs to be improved regarding some mechanical properties such as ductility, fatigue strength, elongation and yield strength. In addition, it has the disadvantage of having different properties in the running direction than in the perpendicular direction.

The phase diagrams of the Cu-Ti system showed that the final Cu(α)-rich solid solution FCC is in equilibrium with a $TiCu_3$ phase characterised by a Pnmm (Orthorhombic Space Group) space group. However, an Au_4Zr type structure has also been reported and this orthorhombic $TiCu_4$ phase is characterised by a Pnma space group (Orthorhombic Space Group). Early studies in the 1970s by Soffa and Laughlin reported that a coherent, metastable, tetragonal $D1_a$ precipitate phase (Ni_4Mo type; I4/m order - tetragonal system)

6

of TiCu$_4$ composition forms during alloy ageing at approximately 600-700 °C (Soffa, 2004). This important finding was verified by Datta and Soffa in an electron microscopic study of alloy series containing 1-4 mass per cent Ti. The resulting metastable mixtures are finely dispersed, two-phase α'+D1$_a$ phases that occur during alloy ageing and produce highly aligned, quasi-periodic microstructures, often referred to as modulated structures.

During prolonged ageing of the alloy at low to moderate ageing temperatures, a coarse lamellar microconstituent composed of the equilibrium phase and the terminal solid solution (α) nucleates and grows, consuming the fine-scale dispersion of coherent/semicoherent D1$_a$ particles. Fig. 1(a) and (b) show two recent versions of the equilibrium diagram and the metastable solution of the D1$_a$ precipitate phase (Yang, 2023). The part of the Cu-Ti equilibrium diagram in Fig. 1(b) shows a polymorphic transformation of the Ti-Cu$_4$ phase and suggests that the tetragonal D1a phase is the stable phase at temperatures below ≈ 500 °C whereas the Au$_4$Zr-type orthorhombic structure is the equilibrium phase at high temperatures. Since D1$_a$ is the ground state for FCC-based alloys, the occurrence of the D1$_a$ structure as the stable phase at low temperatures is not surprising. The Pnma (Au$_4$Zr type) orthorhombic structure is likely to be stabilised at higher temperatures by entropic effects.

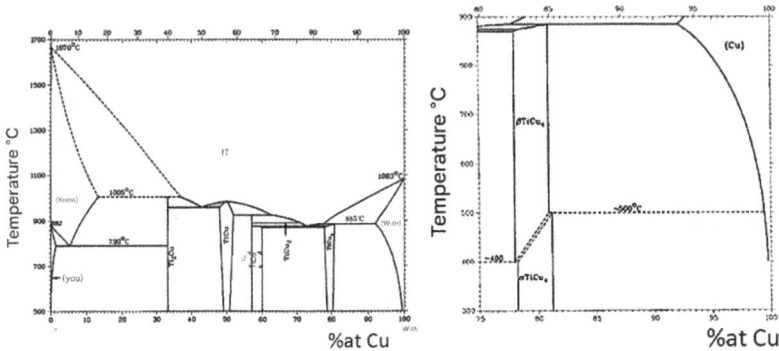

Figure 1. Recent versions of the Cu-Ti phase diagram: (a) Ti Cu4 (orthorhombic; Pnma) shown as the equilibrium phase at temperatures above 500 °C and (b) detailed part of the diagram showing the polymorphic transformation to the Ti Cu4 phase. The x-axis indicates the atomic percentage of Cu (% at)[30].

A significant interest in age-hardenable Cu-Ti alloys was aroused in the 1960s and 1970s, when these alloys were recognised as prototypical binary "side-band" alloys, likely to exhibit spinodal decomposition. Enhanced age-hardening has also received renewed attention, as these precipitation alloys have been found to exhibit exceptional plastic flow behaviour, including abundant deformation necking. The occurrence of spinoidal decomposition in the early stages of precipitate formation led to the formation of an ordered phase. In the production and commercialisation of these alloys, the influence of plastic deformation on ageing behaviour will be considered, as it will be of great importance in the thermo-mechanical processing of these materials in mass production.

Numerous experimental techniques have been used to elucidate the underlying mechanisms that control the formation of the characteristic periodic and aligned two-phase microstructures that occur during ageing, including X-ray and neutron diffraction, electron microscopy and diffraction, and atom probe field ion microscopy (APFIM).

The decomposition of supersaturated Cu-Ti alloys containing 1-6 % Ti is based on a complex interplay of grouping and ordering effects in solid solutions, as well as a subtle synergy between short-range ordering (SRO) and long-range ordering (LRO). In addition, supersaturated CuTi alloys exhibit the well-known 'sideband' phenomena in the diffraction patterns at early stages of decomposition. As emphasised above, the precipitated phase of particular interest in the response to ageing phenomena of these high strength alloys exhibits the $D1_a$ superstructure (Ni_4Mo type; I 4/m order - tetragonal system). A generalised Bragg-Williams-type description of the thermodynamics of the LRO $Cu_4Ti/D1_a$ structure requires a fourth nearest-neighbour atom to stabilise the structure as an ungenerated ground state for FCC-based superstructures, but interactions with the third nearest-neighbour will lead to the emergence of short-range ordering - SRO and long-range ordering in a disordered solid solution at low temperatures. The A4B superstructure is favoured in systems where the first nearest neighbour exchange potential V(1) and the second nearest neighbour exchange potential V(2) are of the same sign, considering only the first and second nearest neighbours in the structural energies of the different FCC-based superstructures [31].

In summary, copper-titanium alloys consist mainly of copper and contain 2 to 6 wt% of Ti, with a preferred Ti content of 3 to 5 wt%. If they contain less than 2 wt% Ti, no appreciable age-hardening effect is to be expected, and the addition of more than 6 wt% Ti does not provide a correspondingly increased age-hardening effect, since the excess above 6 wt% increases the age-hardening effect.

1.6   A brief analysis of the regulations governing the organisation of intervention and rescue operations in industrial installations potentially at risk of toxic/explosive/flammable gas emissions

Employees who work with materials that cause explosive atmospheres or have deflagrating properties are exposed to a wide range of risks. In general, these risks arise from a harmful environment for employees working in such situations. Typical working environments for which employees need to be particularly prepared are highly toxic, low O_2, explosive or highly flammable. In each of these cases there are individual problems that need to be recognised, assessed and dealt with in a specific way, because there is a high possibility of incidents with serious consequences. The assessment of the main standards for the performance of service activities requires the use of new, specialised and intelligent materials to prevent the occurrence of hot sparks causing explosions.

All workers who conduct various activities in potentially explosive atmospheres are exposed to many types of hazards. Generally speaking, these risks are due to the harmful environment in which the workers operate [19,21]. More specifically, they are characterised by individual risks which, if not identified, assessed and properly managed, increase the likelihood of workplace incidents with serious consequences. [32].

Some practical cases encountered at work:

(a) In cases where there is a risk of toxic, asphyxiating, detonating or flammable gases, vapours or dusts being produced during the production/processing protocols, the production company must plan an area on site for the immission and recovery of toxic/poisonous or flammable atmospheres, known as rescue stations [33].

(b) It is accepted to use only rescue centres (stations) approved and supervised by the National Research and Development Institute for Mining Safety and Anti-Explosive Protection - INSEMEX Petroşani - INSEMEX [34].

(c) In order to obtain the prescription referred to in point (b), companies must submit an application to INSEMEX, together with a file containing the following information:

i) information about the rescue station area;

ii) the working team of the company;

iii) a list of the equipment used;

iv) effective methods for reporting and preventing or dealing with emergencies.

Rescue stations will be backed up by operational, checking and operating staff as well as technicians or engineers specialized in aid/repair operations.

Employees working in rescue stations must be trained and authorised by INSEMEX Petroşani.

For these intervention sites, where other specialised units (ISU, SPSU, etc.) may also be involved in the intervention, the level of supply of intervention equipment shall be determined according to the number of intervention teams for the busiest working period, with a 50% reserve. This regulation also applies to companies that accept more than the minimum number of intervention teams required for the busiest shift (I7—2011).

(1) The main requirements for employment in rescue stations are:

a) age between 20 - 65 years for operational staff (> 50 years a medical examination is performed every six months);

b) have a medical and psychological certificate confirming that they can work in contaminated/hazardous or flammable atmospheres.

(2) The control and execution of the intervention steps are performed by technically trained specialists.

Only specially trained people who are not on sick leave or official leave, or who do not have a valid medical record, may participate in interventions to help employees [35].

Employees who have missed the periodic training programme three times in a row will not be accepted for aid trials until they have regained their practical skills and physical fitness, actions to be noted by the aid station coordinator.

(1) The coordinator of the aid station or his/her assistant, together with a medical professional, will facilitate the intervention of the rescuers in the affected area;

(2) If the presence of a medical professional is not possible, the intervention in the affected area of specialised personnel is accepted by the coordinator of the intervention station or his assistant together with the coordinator of the intervention team;

(3) All equipment required for the assistance intervention must be in good working order and must be checked before each intervention.

In the area of intervention, the rescue operations are conducted according to an action protocol drawn up by the coordinator of the rescue team or his assistant according to the proposed action plan [33].

Only workers designated as part of the rescue team may enter the disaster area. In order to keep the equipment of the rescue stations in good working order, routine checks and repairs shall be undertaken at the rescue stations and periodic checks and repairs shall be undertaken at the service units specified by the manufacturer of the equipment. Maintenance, inspection and repair of life-saving appliances, transfer of pressurised gases used in isolating appliances at life-saving stations from units shall be done only by trained and authorised life-saving station mechanics in accordance with Article 5 [36].

Explosion protection is particularly important for safety because explosions endanger the life and health of workers due to the uncontrolled effects of flames and pressure, the presence of noxious reaction products and the consumption of oxygen from the ambient air breathed by workers. In the event of an explosion, workers are exposed to hazards resulting from uncontrolled ignition and pressure phenomena such as heat radiation, flames, shock waves, projection of debris, the presence of harmful reaction products and the oxygen depletion of air essential for breathing. If the formation of *hazardous explosive atmospheres* can be safely prevented, no other measures are necessary.

In many cases it is not possible to avoid explosive atmospheres or ignition sources to a sufficiently safe level. In such cases, measures must be taken to limit the effects of explosions to an acceptable level. So-called non-sparking materials, which are in fact low sparking materials, are often recommended in various failure analysis and remedial action reports.

The criteria for a material to be non-sparking are not clear, nor is the strength of the spark generated by friction quantitatively established. Eliminating frictional sparking from the workplace is only possible by following safe working procedures, minimising the possibility of thermal reaction, controlling tool speed, properly securing loose components, etc. A series of studies and research were conducted to improve the anti-spark properties of some components of equipment used in potentially explosive atmospheres. A particular case would be the gears used in the transmission systems of grinding mills.

1.7　Objectives and prospects

This book proposes to identify solutions to increase the safety of metallic elements operating in potentially explosive environments. This can be achieved by using metallic materials that produce cold sparks or very small amounts of sparks, generally non-ferrous metallic materials, or by using thin ceramic layers as coatings on conventional ferrous metallic elements.

The subject is multidisciplinary in nature, requiring a broad knowledge of materials engineering, physics, chemistry and experimental analysis, allowing for contributions in all areas characterised by the presence of low sparking alloys, methods of deposition of metallic materials and the actual applications of the finished products for the protection of workers and equipment. The complex nature of the experimental determinations results from the general design of the research methodology and the objectives of the study. The experimental research methodology followed a design of experiments based on the behaviour of the materials in service. The aim was to characterise the gear materials and to characterise and control the production process, Fig. 2. The design of the experiments included new analytical techniques that can provide extensive information that can contribute to the desired end product.

The main aim of this work is to identify materials suitable for use in potentially explosive environments and to eliminate the possibility of sparking during the working process. It is proposed to obtain experimental copper-based materials from the Cu-Al-Be and Cu-Ti systems (alloys proposed for the aeronautical industry), which have been tested for the manufacture of gears, to which beryllium will be added in small percentages to reduce the intensity of the sparks generated by the alloy. The alloys are obtained by conventional casting in a controlled atmosphere induction furnace. The experimental alloys were studied using: scanning electron microscopy, VegaTescan LMH II SEM, SE secondary electron detector, 2D and 3D structural analysis, sizing, light intensity variation and other experimental applications of VegaTescan dedicated software on the obtained signal or optical microscopy, chemical analysis by EDAX (Bruker EDAX instrument with PB-ZAF, automatic or element list operating modes or dedicated line, point or map analysis modes), spark spectroscopy (Foundry Master with 3-point averaging), X-ray diffraction (XRD X'PERT PRO MRD). Acquisition software: X'pert Data Collection, and for interpretation X'pert High Score Plus, continuous scan: start angle: 20 and end angle: 120, step: 0.0131303, time: 61, scan speed: 0.05471, number of steps: 7616, 45 KV, 40 mA, (Anode X-ray tube: Cu) and Differential Scanning Calorimetry (DSC Netzsch, Phoenix 204, heating speed 10 K/min, Ar atm.).

Figure 2. Experimental research methodology.

Table 2. shows the main experimental alloys analysed in the study.

Table 2. Labels of the main experimental alloys studied in the study

Experimental alloy	Label	Chemical composition (wt%)	State
CuAlBe	**Alloy 1a**	Cu-8.5wt%Al-0.8wt%Be	Cast/rolled
	Alloy 2a	Cu-8.4wt%Al-0.4wt%Be	Cast / rolled
	Alloy 2b	Cu-10.4wt%Al-0.5wt%Be	Cast / rolled / rolled + annealed
	Alloy 3	Cu-10.3wt%Al-5.2wt%Be	Cast / rolled / rolled + annealed
CuTi	**Cu2Ti**	Cu -2.1 wt% Ti	Cast / rolled
	Cu3Ti	Cu ~ 2.7 wt% Ti	Cast / rolled

The main objectives of this work are:

- developing non-ferrous gearwheel alloys that do not produce hot sparks;
- structural, chemical, mechanical and thermal characterisation of experimental alloys;
- sparking capacity characterisation;
- establishing safe working conditions for metallic gearwheels in potentially explosive environments.

References

[1] J. Bond, Sources of ignition: flammability characteristics of chemicals and chemical products, Butterworth-Heinemann, Oxford, England, 1991.

[2] J.A. Rankine, Workshop processes and materials for mechanical engineering technicians: 2, Penguine Books, England, 1968.

[3] L.G. Britton, R.J. Willey, Pros and cons of non-sparking tools, Process Saf. Prog. 41 (2022) 1-10. https://doi.org/10.1002/prs.12347

[4] S. Kalyanam, P.N. Sankaran, Accidental histories in composite propellant processing and safety measures, Report No. ISRO-SHAR-TR-08-059-91, Shar Centre, Sriharikota, India, 1991.

[5] L.D. Huang, C. Yan, Y. Zhou, W. Yan, Multi-directional forging and aging treatment effects on friction and wear characterization of aluminium-bronze alloy, Mater. Charact. 167 (2020) 110511. https://doi.org/10.1016/j.matchar.2020.110511

[6] R.M. Bruce, M. Odin, Beryllium and beryllium compounds, UNEP/ILO/WHO Publication, Geneva, 2001.

[7] P.J. Earley, B.L. Swope, M.A. Colvin, G. Rosen, P.-F. Wang, J. Carilli, I. Rivera-Duarte, Estimates of environmental loading from copper alloy materials, J. Bioadhesion Biofilm Res. 36 (2020) 1-10. https://doi.org/10.1080/08927014.2019.1708334

[8] R.K. Singh, S. Murigendrappa, S. Kattimani, Investigation on properties of shape memory alloy wire of Cu-Al-Be doped with zirconium, J. Mater. Eng. Perform. 29 (2020) 11-15. https://doi.org/10.1007/s11665-020-05233-7

[9] A.C. Canan, K. Oktay, N. Ünlü, I. Özkul, Study on basic characteristics of CuAlBe shape memory alloy, Braz. J. Phys. 51 (2020) 13-18. https://doi.org/10.1007/s13538-020-00823-1

[10] Y. Jia, L. Li, X. Liao, Z. Xin, Y. Li, Y. Zhang, Z. Li, Y. Pang, J. Yi, Effects of thermo-mechanical processing on microstructure and properties of Cu-0.9Hf alloy, J. Alloys Compd. 1003 (2024) 175574. https://doi.org/10.1016/j.jallcom.2024.175574

[11] M. Pedrosa, D. Silva, I. Brito, R. Alves, R. Caluête, R. Gomes, D. Oliveira, Effects

of hot rolling on the microstructure, thermal and mechanical properties of CuAlBeNbNi shape memory alloy, Thermochim. Acta 711 (2022) 179188. https://doi.org/10.1016/j.tca.2022.179188

[12] HG 752/2004 privind condiţiile pentru introducerea pe piaţă a echipamentelor şi sistemelor protectoare în atmosfere explozive (actualizată 2023).

[13] P. Lerena, G. Suter, Tools to assess the explosion risks in the chemical, pharmaceutical and food industry, Health Environ. Res. Online 19 (2010) 243-248.

[14] NP 018-97 (2023) Proiectarea, execuţia şi exploatarea punctelor de desfacere a buteliilor cu GPL.

[15] Directive 2014/34/EU of the European Parliament and of the Council, 2014.

[16] K. Sekiguchi, F. Yasui, E. Fujii, Capturing of gaseous and particulate pollutants into liquid phase by a water/oil column using microbubbles, Chemosphere 256 (2020) 126996. https://doi.org/10.1016/j.chemosphere.2020.126996

[17] R.L. Quirino, L. Richa, A. Petrissans, P.R. Teixeira, G. Durrell, A. Hulette, B. Colin, M. Petrissans, Comparative study of atmosphere effect on wood torrefaction, Fibers 11 (2023) 27. https://doi.org/10.3390/fib11030027

[18] D.G.P.S.I.-2004 Dispoziţii generale privind reducerea riscurilor de incendiu generate de încărcări electrostatice.

[19] Directive 89/391/EEC şi Legea nr. 319/2006 privind securitatea şi sănătatea în muncă.

[20] D.M. Candidoa, G.V. Ferreira de Oliveira, D. Brito, I.C.A. Evaristo Caluête, B.H. da Silva Andrade, D.G.L. Cavalcante, Mater. Res. 23 (2020) 1-5. https://doi.org/10.1590/1980-5373-mr-2019-0542

[21] Directive 1999/92/EC of the European Parliament and of the Council, 1999.

[22] T. Piotrowski, How to prepare an explosion protection document, Institute of Industrial Organic Chemistry, Warsaw.

[23] SR EN 1127-1:2011 Atmosfere explozive. Prevenirea şi protecţia la explozii.

[24] D.-C. Darabont, R.I. Moraru, A.E. Antonov, C. Bejinariu, Managing new and emerging risks in the context of ISO 45001 standard, Qual. Access Success 18 (2017) 11-14.

[25] J.T. Al-Haidary, A.M. Mustafa, A.A. Hamza, Effect of heat treatment of Cu-Al-Be shape memory alloy on microstructure, shape memory effect and hardness, J. Mater. Sci. Eng. 6 (2017) 6-10. https://doi.org/10.4172/2169-0022.1000398

[26] N. Nagel, Beryllium and copper-beryllium alloys, ChemBioEng Rev. 5 (2018) 30-33. https://doi.org/10.1002/cben.201700016

[27] H. Zhang, Ya. Jiang, J. Xie, Yo. Li, L. Yue, Precipitation behavior, microstructure and properties of aged Cu-1.7 wt% Be alloy, J. Alloys Compd. 773 (2019) 1121-1130. https://doi.org/10.1016/j.jallcom.2018.09.296

[28] Z. Cui, L. Huang, X. Meng, Q. Lei, Z. Xiao, Z. Li, Research progress of ultrahigh-strength copper-titanium alloys, Metall. Eng. 7 (2020) 121-129. https://doi.org/10.12677/MEng.2020.73018

[29] P. Mahmoudi, M.R. Akbarpour, H.B. Lakeh, F. Jing, M.R. Hadidi, B. Akhavan, Antibacterial Ti-Cu implants: a critical review on mechanisms of action, Mater. Today Bio 17 (2022) 1-15. https://doi.org/10.1016/j.mtbio.2022.100447

[30] H. Yang, B. Zeng, H. Wang, H. Jin, C. Zhou, Phase equilibria and thermodynamic re-assessment of the Cu-Ti system, CALPHAD 82 (2023) 102594. https://doi.org/10.1016/j.calphad.2023.102594

[31] B. Rouxel, C. Cayron, J. Bornand, P. Sanders, R.E. Logé, Micro-addition of Fe in highly alloyed Cu-Ti alloys to improve both formability and strength, Mater. Des. 213 (2022) 110340. https://doi.org/10.1016/j.matdes.2021.110340

[32] Directive 2014/34/EU of the European Parliament and of the Council, 2014.

[33] NP 099-04 Normativ pentru instalații electrice în zone cu pericol de explozie, 2005.

[34] Normativ pentru stații de distribuție carburanți la autovehicule, 2005.

[35] SR EN 1127-1:2011 Atmosfere explozive. Prevenirea și protecția la explozii.

[36] HSG 103, Manual explosion protection: safe handling of combustible dusts, HSE Books, 2003.

CHAPTER 2

Materials and Experimental Procedures used in the Analysis of Anti- Sparking Alloys

Romeo-Gabriel CHELARIU[1], Ramona Cimpoeşu[1]*,Gabriel-Dragoş VASILESCU[2],Costică Bejinariu[1,3]

[1]Faculty of Materials Science and Engineering, "Gheorghe Asachi" Technical University of Iasi, 67 Dimitrie Mangeron Street, 700050 Iasi, Romania

[2]National Institute for Research and Development in Mine Safety and Protection to Explosion— INSEMEX, 332047 Petrosani, Romania

[3]Academy of Romanian Scientists, Ilfov 3, 050044 Bucharest, Romania

ramona.cimpoesu@academic.tuiasi.ro

Abstract

In this chapter, the materials investigated and the experimental procedures applied for the characterization and analysis of anti-sparking alloys are described in detail. The section provides an overview of the preparation conditions, testing environments, and analytical techniques employed to evaluate their structural, chemical, and electrochemical behavior. Several cast and rolled CuTi alloys have been proposed for anti-spark applications as replacement materials for conventional CuBe alloys. The CuTi system has been thermodynamically evaluated in terms of experimental equilibrium phases and their thermochemical properties.

Keywords

CuTi Alloys, CuAlBe, CuTi, Chemical Composition; Mechanical Analysis, Electrochemical Behavior

2.1 Chemical composition determination of CuTi alloys using ThermoCalc

The main solution phases are: Liquid (L), fcc-A_1 (Cu), bcc-A_2 (β-Ti), and hcp-A_3 (αTi) has been considered substitutional. Intermetallic compounds that can be formed: $TiCu_2$, Ti_2Cu_3, Ti_3Cu_4 and $CuTi_2$ with negligible solubility were described as lines with the formula Cu_pTi_q, while β- $TiCu_4$ and $CuTi$ with higher solubility were modelled with the formula $(Cu,Ti)_r(Cu,Ti)_s$. A set of reliable thermodynamic parameters of the Cu-Ti system has been obtained. The calculated results are in good agreement with the experimental results reported in the literature. The formation of the $CuTi_2$ compound was observed and the $CuTi_3$ compound was not identified with the equipment in the cast samples. The temperatures of nine invariant reactions have been determined.

New Shock-Resistant Materials for Work Equipment used in Potentially Explosive Atmospheres
Materials Research Foundations **186** (2026)   https://doi.org/10.21741/9781644903872

The thermodynamic and kinetic databases within the ThermoCalc or CalPHAD method are useful for the design and development of alloys in general and Cu and Ti in particular [1], and reliable thermodynamic parameters of the relevant systems can be evaluated first if required. Wang et al. have performed thermodynamic modelling of the Cu-Ti system and Fig. 1 shows the calculated phase diagram together with the enthalpies of liquid phase mixing at 1450°C and the enthalpies of intermetallic formation at 25 °C [2].

Some of the calculated results of Wang et al. have generated some debates related to: (1) invariant reactions, especially in the Cu-rich corner, remain questionable; (2) peritectic or congruent formation of CuTi2 is controversial; (3) the liquid phase mixing enthalpies at 1300 and 1600 °C from Turchanin et al. [3] are not taken into account.

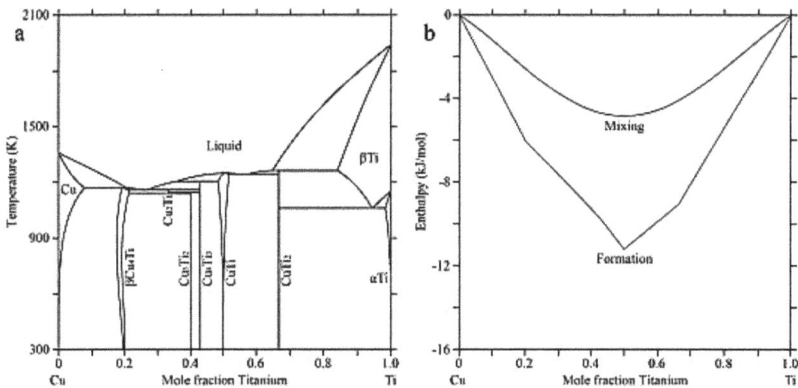

Figure 1. (a) Calculated Cu-Ti phase diagram; (b) Calculated mixing enthalpies for the liquid phase at 1450 °C and enthalpies of intermetallic formation at 25 °C of the Cu-Ti system [2].

In the light of experimental information prior to 1983 [4], Murray critically revised the phase equilibrium of the Cu-Ti system (Murray, 1983). The Cu-Ti stable phase diagram included six intermetallic compounds: β- TiCu4, TiCu2, Ti2Cu3, Ti3Cu4, CuTi and CuTi2 and four phases of the solution Liquid (L), fcc-A1 (Cu), bcc-A2 (βTi) and hcp-A3 (αTi). Subsequently, Eremenko et al. investigated the homogeneity range and congruent melting temperature of CuTi by metallography, X-ray diffraction (XRD) and differential thermal analysis (DTA) [5]. Taguchi et al. analysed the reaction diffusion between Cu and Ti from 690 to 870 °C. The main crystal structures and thermodynamic models of all phases in the Cu-Ti system, according to the literature, are given in Table 1 [6].

The phase relations in the region of the equilibrium diagram with higher Ti content for different pressures were determined using electron probe microanalysis (EPMA) by Yamane et al. They found that the eutectoid temperature (βTi ↔ CuTi2 + αTi) decreases with increasing pressure, while the peritectic temperature (L + βTi ↔ CuTi2) increases [7]. Nagarjuna and Sarma measured the electrical resistivities and lattice parameters of four

Cu-Ti alloys heat treated by solution quenching at 900 °C, from which the phase boundary between Cu and Cu + β- TiCu₄ could be deduced [8].

Table 1. Crystal structures and thermodynamic models of all phases in the Cu-Ti system [9].

Phase	Designation of structural report	The Pearson symbol	Space group	Prototype	Thermodynamic model
Liquid	-	-	-	-	$(Cu,Ti)_1$
Cu	A1	cF4	Fm3m	Cu	$(Cu,Ti)_1 Va_1$
βCu₄Ti	–	oP20	Pnma	Au₄Zr	$(Cu,Ti)_4(Cu,Ti)_1$
αCu4Ti[a]	D1ₐ	tI10	I4/m	MoNi₄	–
Cu₂Ti	–	oC12	Amm2	Au₂V	Cu_2Ti
Cu₃Ti₂	–	tP10	P4/nmm	Cu₃Ti₂	Cu_3Ti_2
Cu₄Ti₃	–	tI14	I4/mmm	Cu₄Ti₃	Cu_4Ti_3
CuTi	B11	tP4	P4/nmm	γCuTi	$(Cu,Ti)_1(Cu,Ti)_1$
CuTi₂	C11ᵦ	tI6	I4/mmm	MoSi₂	$CuTi_2$
CuTi₃[a]	–	oP28	Pbam	–	–
βTi	A2	cI2	Im3m	W	$(Cu,Ti)_1 Va_3$
αTi	A3	hP2	P6₃/mmc	Mg	$(Cu,Ti)_1 Va_{0.5}$

[a] Metastable phase.

Zhan et al. found the existence of Cu-Ti₃ compound by scanning electron microscopy (SEM) performed simultaneously with energy dispersive X-ray spectroscopy (EDX), XRD and DTA, but its crystal structure and phase stability were still undefined [10]. Laik et al. used six diffusion couples to study the diffusion characteristics of Cu-Ti system from 720 to 870 °C. β- TiCu₄, Ti₃Cu₄, CuTi and CuTi₂, were identified, but Ti₂Cu₃ was not found by EPMA analysis [11]. Table 1 shows the details of the crystal data for all phases.

The mixing enthalpies of the liquid phase have been studied by high-temperature reaction calorimetry, the Knudsen effusion method, and differential calorimetry [12]. Arita et al. used a Sieverts equipment to determine the enthalpies of βCu₄Ti and CuTi formation at 500 °C [13.] Colinet et al. measured the enthalpies of βCu₄Ti, Cu₃Ti₂, Cu₄Ti₃, CuTi and CuTi₂ phases and compounds formation at 25 °C by solution calorimetry [14]. Over the past decade, numerous studies have been reported to determine the enthalpies of formation of various intermetallic compounds using first-principles thermodynamic calculations [15]. In the recent work of Turchanin et al, CuTi₃ was treated as a metastable phase [3].

The analysis parameters used were based on the mass percentage, mole fraction and Gibbs energy characteristic of the Cu-Ti system. Two calculation systems have been developed in the ThermoCalc alloy phase analysis program, Fig. 2 a) for the mass percentage and Fig. 2 b) for the Gibbs free energy analysis (thermodynamic value of the system that provides the energy required for the transformation of the Cu-Ti system into Cu-Ti alloy).

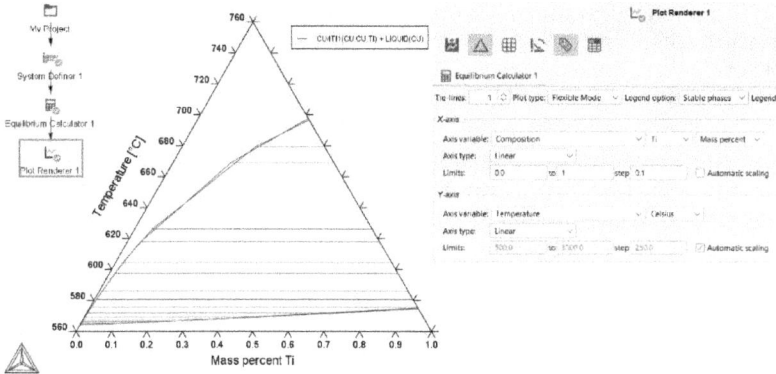

(a)

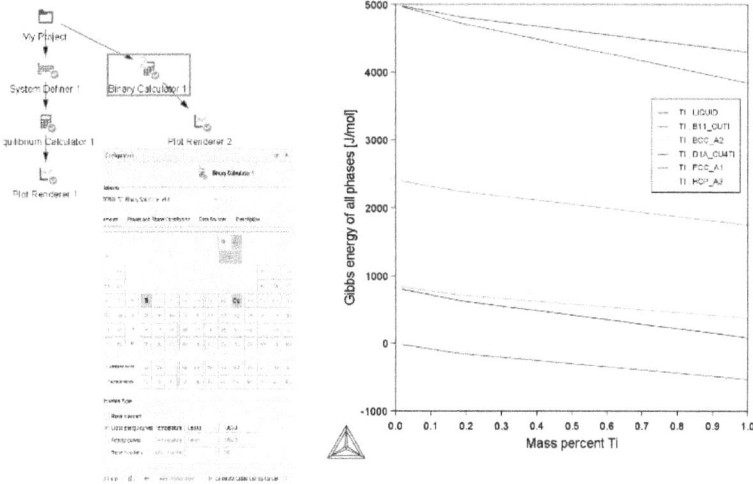

(b)

Figure 2. Parameter settings of the CuTi system analysis in the specialized ThermoCalc software (a) the variation of the possibility of forming phases and compounds as a function of the percentage variation of titanium and (b) the variation of the possibility of forming phases and compounds as a function of the percentage variation of titanium and Gibbs energy.

The software is used to determine whether a reaction in a thermodynamic process is spontaneous. Simply put, spontaneous reactions are those that occur naturally during the heating or cooling , while non-spontaneous reactions are those that do not. When the term "naturally" is used, it refers to a reaction that can occur in a given system without the influence or input of a net energy flow that is freely available from the environment.

As part of the analysis method using the ThermoCalc software, several thermodynamic calculations of the Cu-Ti system were performed, which were confirmed by the results presented in the literature [3,16]. All stable phases as well as the very high solubility of $TiCu_4$ and CuTi compounds were accounted for and a complete thermodynamic modeling of the system was performed, Fig. 2 (a) confirms the results presented by Kumar et al. obtained with the CALPHAD software [16]. The plots in Fig. 3 were obtained using the model in Fig. 2 (a), and those in Fig. 4 and Fig.5 were obtained using the model in Fig. 2 (b). The legend to Fig. 2 (a) lists all the possibilities suggested by ThermoCalc and the temperature conditions under which they can be formed, beginning with the mole fraction of Ti.

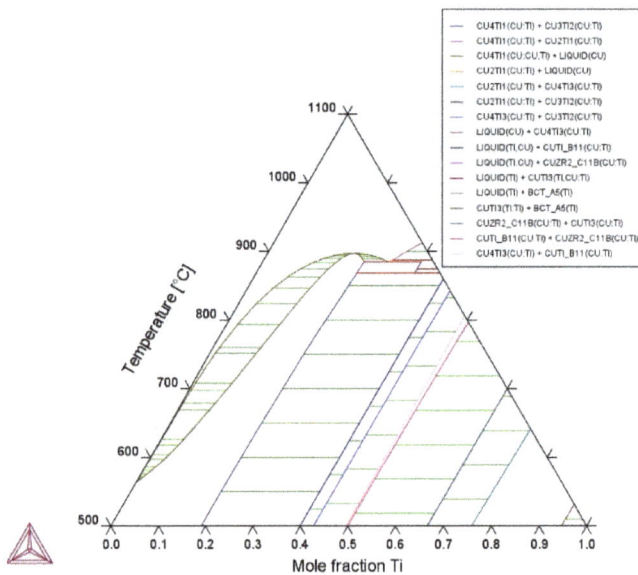

(a)

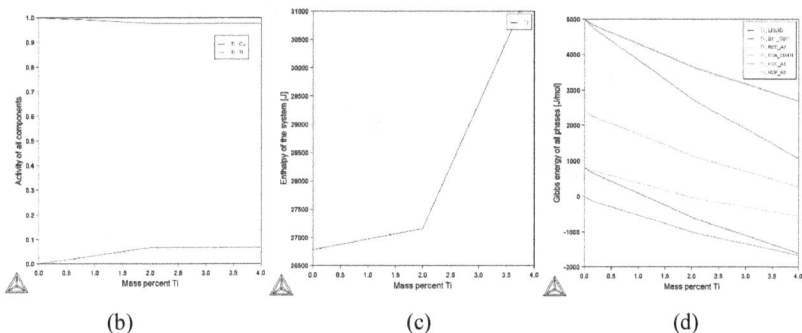

(b)   (c)   (d)

Figure 3. Diagrams generated by ThermoCalc for the CuTi system: (a) Diagram of the compounds that can be formed in the Cu-Ti system as a function of the temperature and the molar fraction of titanium varying from 0 to 1, (b) the activity variation of the components of the Cu-4wt%Ti system as a function of the mass percentage of Ti, (c) the enthalpy variation of the Cu-4wt%Ti system as a function of the mass percentage of Ti, and (d) the Gibbs free energy variation of the Cu-4wt%Ti system as a function of the mass percentage of Ti.

Within the Cu-Ti system, the phases are generally formed in the following order: $TiCu_4$, $TiCu_2$, Ti_3Cu_4, CuTi and $CuTi_2$, from a very high mass percentage of Cu, the left corner of the diagram in Fig. 2(a), to the Ti-rich end, and their sizes are generally less than 5µm, although diffusion is allowed for an exposure time of at least 60 minutes [17]. For the applications proposed in this work, i.e. a non-sparking alloy, the focus was on a CuTi alloy with a Ti mass fraction of up to 4%. The chemical composition was chosen based on literature review and Gibbs free energy of the phases, Fig. 3 (d).

From the analysis of the variations of the components' activity, the free energy of the system and the Gibbs energy of the specific phases, shown in Fig. 4, it is proposed to realize a copper alloy with 2-4 wt% Ti, because this chemical composition is in a good range of phase activity, Fig. 4 (a), a very high energy of the alloy system, Fig. 4 (b) and at a sufficiently high level of Gibbs free energy for the formation of compounds that will favor the mechanical properties of the alloy such as its hardness.

The phases are formed along the diffusion gradient and their growth is controlled by Cu and Ti interdiffusion in each other's matrix, Fig. 4(a). Their morphology is fairly uniform and is further determined by the diffusion pathways [18]. However, due to the liquid Cu present, the reaction rate is several times higher than in a diffusion couple, which may allow the phases to form into significantly larger grains [6]. In contrast to the nearly parallel boundaries, the corresponding morphology of the present phases does not follow a specific pattern, as the $CuTi_2$ $CuTi_2$ phases are almost always surrounded by CuTi and Ti_3Cu_4 CuTi counterparts.

21

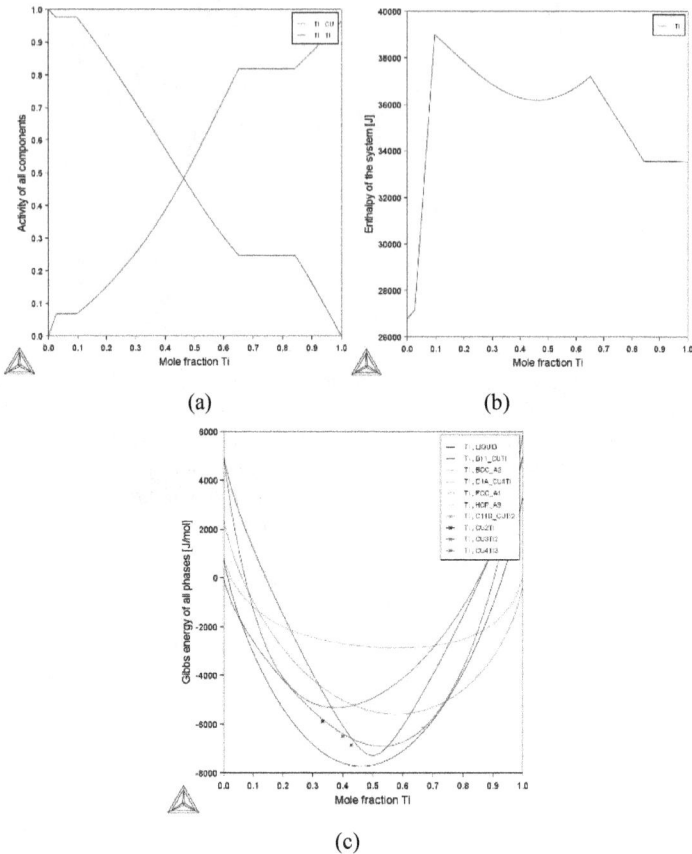

(a)   (b)

(c)

Figure 4. The simulated variations of the Cu-Ti system as a function of the Ti mole fraction (a) activity of the Cu and Ti components, (b) enthalpy of the CuTi system, and (c) Gibbs energy characteristic of the main components of the CuTi system.

The formation of an alloy with the following phases in the cast state is considered for the proposed alloy (Cu-(3-4)wt%Ti): Cu+β-TiCu$_4$.

2.2   Experimental CuAlBe and CuTi based alloy production

To obtain aluminium bronzes, the casting process can be started either from pure metallic elements or from various types of aluminium bronzes as ingots or scrap. The basic metal batch in the case of aluminium bronzes can be composed of the following: cathodic copper, (various labels); technical aluminium, SREN 576 - 2004 and SREN 1676 - 1998; technical aluminium scrap from the electrical or other sectors: bars, wires, sheets, etc; Cu-Al pre-

22

alloy ingots. The production of aluminium bronzes from elements requires the completion of the following operations:

(a) The processing unit must be cleaned;

(b) The processing unit must be heated until its refractory lining has reached a temperature of 850-950 °C (bright red);

(c) A quantity of coating flux must be poured into the lower part of the crucible, which is deemed sufficient to cover the surface of the metal bath;

(d) The pieces of cathodic copper or cathodic copper scrap must be placed into the unit;

(e) The charge must be filled with smaller pieces of metallic material not originally introduced.

(f) The melting process should be continued until the temperature of the metal bath reaches 1150-1200°C, upon completion of the protective flow layer;

(g) The furnace should be switched off and the deoxidising sample should be poured;

(h) The course of processing should be decided upon, depending on the sample configuration;

(i) The bath should be allowed to settle and cool to a temperature of 1120-1140°C, in order to introduce the aluminium;

(j) The metallic or pre-aluminium alloy should be introduced into the bath, which should be preheated to 150-180°C, under the slag layer. The following text is intended to provide a comprehensive overview of the subject matter;

k) The bath should be homogenised with a steel bar in order to avoid "breaking" the slag layer or drawing it into the melt;

l) The other alloying elements should be added as pre-alloys or in solid form and the bath homogenised;

m) The melt should be superheated to the pouring temperature, which is selected according to the components to be cast, and should be maintained at this temperature for approximately 5-10 minutes.

n) The furnace should be switched off and the slag removed.

o) The bronze should be poured into the preheated red-hot casting vessel.

p) It should be poured from a height as low as possible into the mould.

Utilising the primary CuBe (4%wtBe) alloy and high-purity Al, an CuAlBe alloy (~10wt%Al and ~2wt%Be) was obtained in an electric furnace at 1150°C using a ceramic crucible, Fig. 5 (a). The resultant homogeneous ingots, measuring 10 mm in diameter and 100 mm in length, were obtained in a metallic mold. Subsequent heat treatment at 800°C for ten minutes and hot rolling using a rolling mill were then applied to the CuAlBe ingots (see Fig. 5(b)). The degree of reduction was ~10% of the original diameter until plates with a length of 16-18 mm and a thickness of 2-3 mm were obtained. It should be noted that each reduction step was preceded by heating the material to 800 °C.

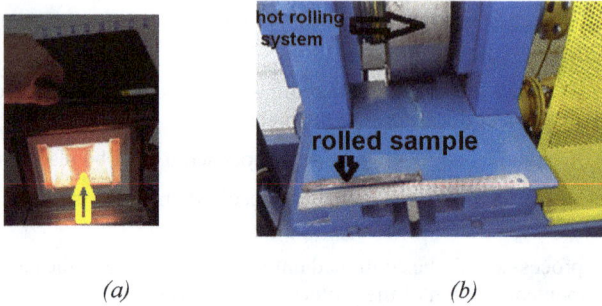

(a)   (b)

Figure 5. (a) Processing equipment, specifically an electric furnace, and (b) a laboratory hot rolling system for CuAlBe.

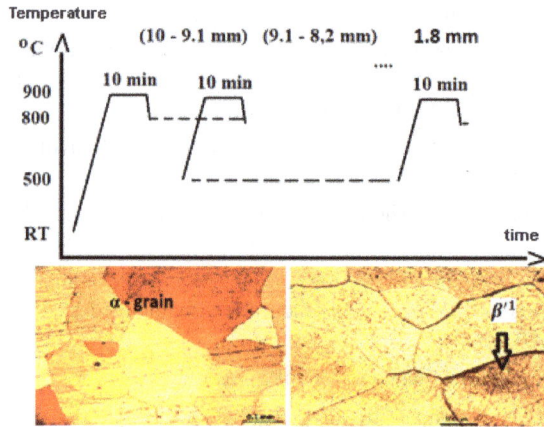

a)

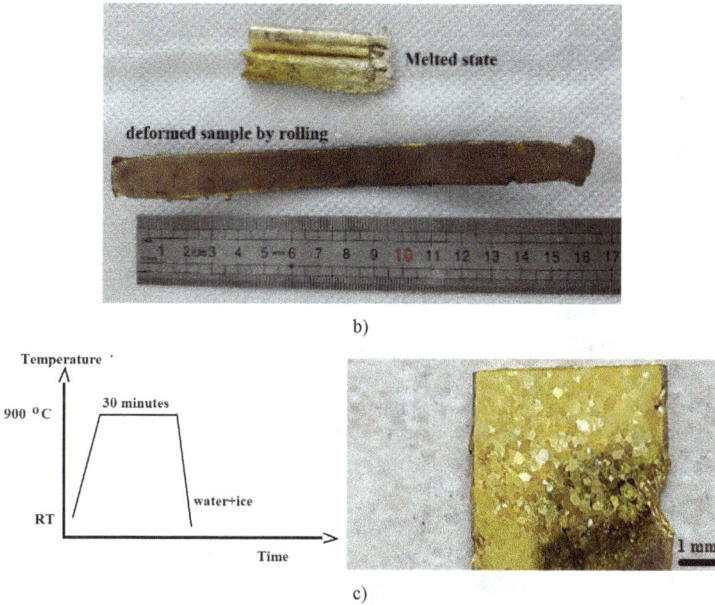

Figure 6. The schematic diagram illustrates the CuAlBe hot rolling process and microstructure (optical microscopy) before and after deformation in a), the material's appearance before (as an ingot) and after (as a plate) hot rolling in b), and the final heat treatment diagram of the rolled material in c).

It is observed for the sample with more Be a polycrystalline structure of the alloy at room temperature of austenite type and after rolling narrow zones of stress induced martensite. Structural analysis showed an abnormal grain growth by rolling, also reported in other literature [19], an orientation of the structure in the rolling direction and the appearance of $\beta'1$ stress-induced martensite, Fig. 2.

Cu-Ti alloys were obtained by melting pure copper and high-purity titanium (99.99%) in an electric furnace at 1100°C. The chemical composition of the alloys designated as Alloy 1 and Alloy 2 is as follows: Cu: ~98 wt% and Ti: ~2 wt%. The ingots, with a diameter of 10 mm and a length of 100 mm, were subjected to hot rolling using the equipment shown in Fig. 5 (b), namely a laboratory furnace with electric resistance heating and rolling equipment. The resultant plates were 3 mm thick and 160 mm long. No longitudinal cracks were observed, and the integrity of the material remained unaffected by the thermo-mechanical process applied.

Subsequent to the hot rolling process, the materials underwent a heat treatment that involved a water-and-ice quenching procedure, as illustrated in Fig. 6(c). The subsequent heat treatment, designated as solution quenching, was implemented to facilitate the formation of the martensitic phase in CuAlBe alloys. Additionally, an aging heat treatment was applied to Cu-Ti alloys.

2.3   Mechanical properties analysis

The mechanical properties of the materials, in the form of plates, were determined using a laboratory equipment Tribometer: CETR-UMT2 [20], as illustrated in Fig. 7.

The following steps are required to perform the indentation test:

- Microindentation tests can be performed using a Rockwell indenter, Vickers indenter, diamond tip (tip radius 5 or 12 μm) with a capacitance sensor. The schematic of the microindentation test using a Vickers indenter is shown in Fig. 7.

- The sample should be mounted on the sample holder provided. It is important to note that a variety of mounting techniques can be employed for mounting the sample on the sample holder. In the case of a sample that is heavy, it can be placed on the holder under its own weight. Conversely, if the sample is very flat and light, it can be adhered to the holder with wax.

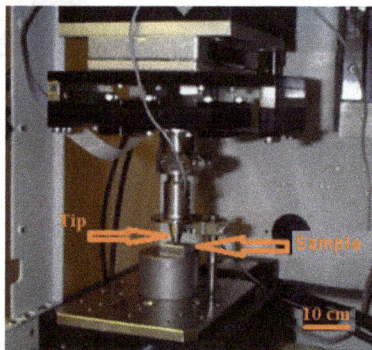

Figure 7. The UMT microindentation test setup.

-The lower linear unit (y-axis) should be moved towards the indenter using the keyboard. For macro positioning, the Alt < or Alt > keys can be used, while the Ctrl < or Ctrl > keys are used for fine positioning. It is imperative to ensure that no obstacles are present that could hinder the process. Continue to move the linear unit until the sample is positioned directly beneath the indenter;

-Utilize the slider to adjust the indenter's position in the X direction, and employ the capacity sensor holder, which can be rotated around the vertical shaft when the screw holding it is loose, to position the indenter in close proximity to the capacity sensor. It is imperative that the capacitance sensor be in close proximity to the indenter to ensure optimal results. The capacitance sensor should be positioned below the reference disk;

-After x positioning is complete, the indenter should be slowly lowered until it is very close to the sample surface (use the Ctrl + down arrow keys). It is imperative to ensure that during this movement, the indenter or reference disk does not make contact with the sensor/capacity holder. If the capacity sensor holder/capacity sensor is hindering the process, it is necessary to lower the capacity sensor holder;

- When the indenter is in close proximity to the surface (or just barely touching the surface), the capacitance sensor holder should be elevated (after loosening the screw). The sensor should be adjusted to a range of 220-230 microns at the reference disk. This adjustment should be made with the aid of the capacitance graph, which will display the distance, and a 200-micron reed that is included with the equipment. When the sensor is out of range, it will show a full scale of 254 microns. The configuration of the setup is depicted schematically in Fig. 8.

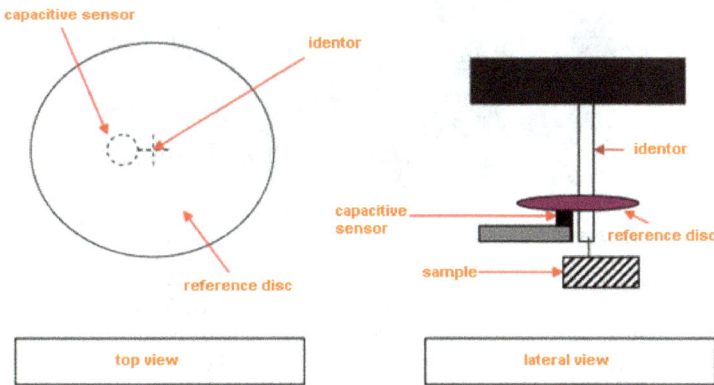

Figure 8 Schematic of Micro Indentation Test Performed on UMT Equipment.

The micro-scratch test protocol is delineated in Fig. 9. The test will consist of a 3-step sequence that will perform multiple scratches, with a separation distance of 1 mm between two scratches and a scratch length of 5 mm. It is noteworthy that the micro-scratch test can be executed with either a constant or a linearly increasing load. The script parameters for performing such multiple micro-scratches using a linear drive for sample displacement during scratching and the use of the slider for positioning are given below.

For calibrating, micro-scratch tests were performed with a constant load of 10 N and a linearly increasing load up to 10 N with a 200 N (20 kg) force sensor installed in the UMT, the force unit is in Newtons. In the event that the UMT option file contains a different unit of force (Fz and Fx), the force values in the equipment program must be altered accordingly. It is imperative to ensure the provision of adequate sensor protection within the Options menu. This can be accomplished by navigating to the Options tab, followed by selecting Edit, then selecting Channels, and finally, enabling the "warn if absolute value exceeds..." and "abort if absolute value exceeds..." options.

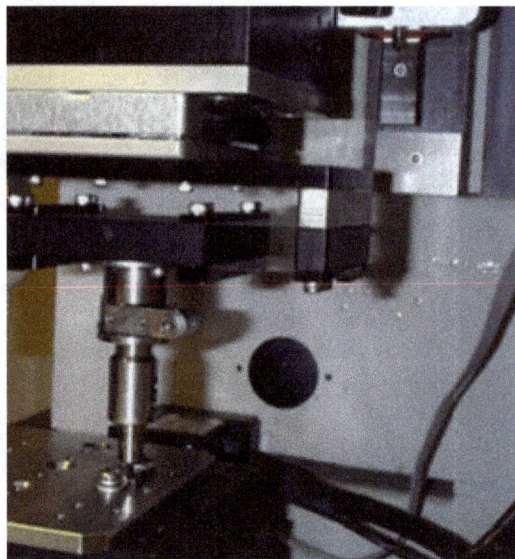

Figure 9. Schematic configuration of the UMT Micro-Scratch Test.

The force values for the warning and cancel actions should be set to slightly less than full scale. It is imperative to reiterate the sensor protection check for the Fx channel. These checks are essential for safeguarding the force sensor against potential damage. Initially, the aforementioned script should be executed once, and the tst file should be opened in UMT Viewer. The force profile (Fz) should be examined to ensure it is stable (linear in case of linear growth) and has reached the desired maximum force. In the event that the aforementioned conditions are not met, it is necessary to repeat the procedure after making adjustments to the engagement parameters. It is acknowledged that engagement parameters may vary slightly from system to system due to the variation in stiffness of the system. In the case of multiple scratches, it is imperative to repeat the tests as many times as necessary at the beginning of the test.

For the experimental CuAlBe samples, the tests were performed on parallel-sided samples with dimensions of 90x18x8 millimeters. The microhardness tests (five on each sample at a distance of 0.30 mm from each other) were performed with a Rockwell diamond tip (120°). The technical specifications of the Rockwell tip indenter are as follows: radius: 210 ± 7 μm; angle: 120° ± 0.25°; and a standard deviation of ± 1.5 μm. Scratch tests were performed to determine the friction coefficient of the obtained alloys and to observe the behavior of the structure (grains and grain boundaries) under an external mechanical force. The experiment was conducted by applying an increasing force from 1 to 25 Newton (N) over a distance of 70 millimeters (mm) with a velocity of 1 millimeter per second (mm/s) on the surface of the etched sample. The results were interpreted using Test Viewer software. Samples of the cast and rolled Cu-Ti material, in the form of plates, were utilized for the determination of several surface characteristics using a tribometer: Cetr-Umt2 [21]. The experiments were conducted on parallel-sided specimens with dimensions of $50\times20\times2$ mm. The scratch tests were performed on the polished surface of the samples, specifically focusing on the grains and grain boundaries. The experiment was conducted by applying an increasing force from 1 to 25 N over a distance of 10 mm at a speed of 1 mm/s on the etched sample surface.

2.4   General test procedure for non-sparking materials

The procedure delineates the protocol for the acceptance testing of spark protection materials intended for utilization in potentially explosive atmospheres within the test group for the purpose of verifying the performance of materials and equipment.

The procedure outlined herein is applicable within the context of laboratories engaged in the acceptance testing of spark-protection materials intended for utilization in potentially explosive atmospheres.

The normative references employed in the execution of this procedure have been extracted from the following standards: Romanian Standard STAS 10449-86 ("Electrical equipment for potentially explosive atmospheres - Impact and friction tests"); SR EN ISO/IEC 17025:2018, General requirements for the competence of testing and calibration laboratories [22].

The primary terminology employed in this test procedure is as follows: *friction spark* - a glowing particle of a material that is detached as a result of friction (or impact); *non-hazardous spark* - a friction (or impact) spark that does not ignite the explosive mixture under prescribed test conditions; *enclosure* - a protective covering or component part of equipment that is in contact with the explosive mixture in the working area; *friction (or impact) spark protective coverings* - protective coverings that ensure the safe operation of equipment in the event of the generation of friction (or impact) sparks.

In Romania, two general procedures are in place for the testing of non-sparking materials. These procedures are presented in the following.

The friction test: is a vital component of the evaluation process for non-sparking protective materials intended for utilisation in potentially explosive atmospheres. This test, conducted under controlled laboratory conditions, aims to replicate the process of spark formation caused by accidental or technological friction in explosive gas mixtures.

The impact test: is performed under laboratory conditions and consists of the simulation on special stands of the formation of sparks generated by accidental or technological impacts.

The simulation is conducted on a dedicated test stand, where an impact is generated between a plate tilted at 350° to the vertical and the specimen to be tested.

The configuration of the weight is selected based on the designated function of the material utilised in the construction of the equipment. The parameters m and h, which are determined by mutual agreement between the parties or in accordance with the applicable regulations, are specific to the impact test.

In the context of material testing of Group I and II A equipment, an explosive mixture consisting of air with 6.5% CH_4 is to be employed. For material testing, an explosive mixture of air with 10% H_2 is to be used. Friction tests, impact tests, or both, may be conducted to ascertain whether hot sparks capable of causing ignition may occur when utilising a particular material in an explosive environment under specific conditions. In the context of gears, the sparking potential is determined by the friction process.

In order to execute the friction test, as illustrated schematically in Fig. 10, the following parts are required: a special stand equipped with a rotating disc driven by an electric motor; a test chamber measuring $1000 \times 1000 \times 1000$ mm; samples of material pairs with the dimensions indicated in Fig. 11 a) and a movable support on which the plate is fixed, which can perform a translation movement parallel to the axis of the rotating disk; the sample to be tested having the rounded friction surface, which is fixed on the rotating disk; an automatic installation for creating an explosive mixture, consisting of the dosing installation and the oxygen analyzer.

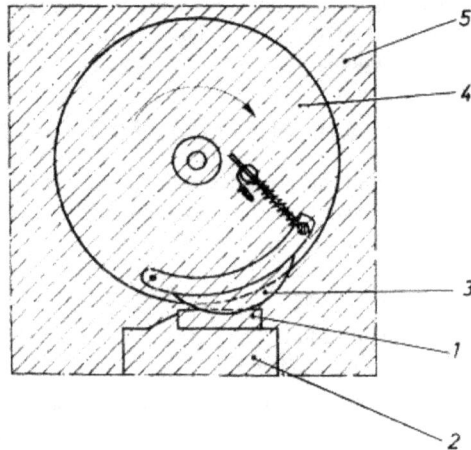

Figure 10. Specific friction test stand; Legend: 1 - test plate; 2 - movable support; 3 - test specimen; 4 - rotating disk and 5 - test medium (explosive mixture).

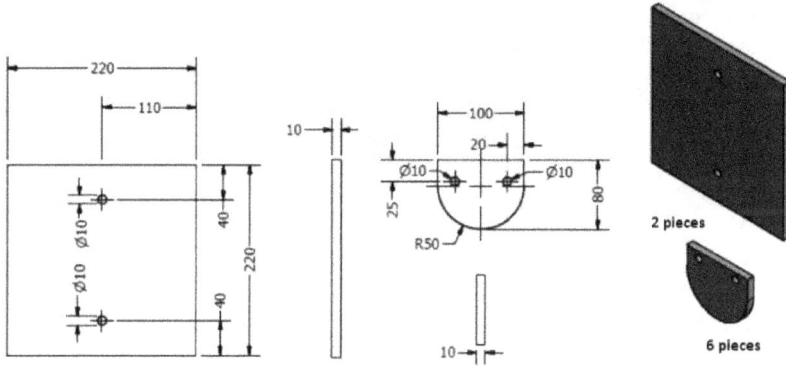

(a)

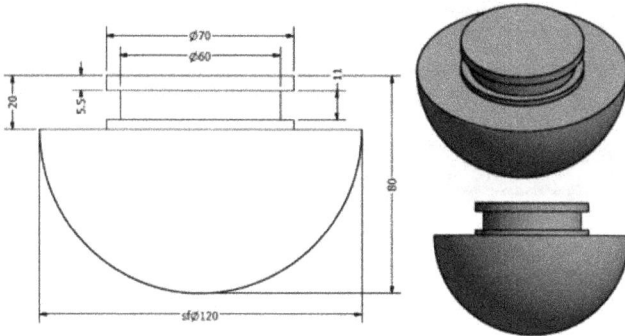

(b)

Figure 11. Metal test specimens (test specimens) fabricated from pairs of materials to be tested, dimensions in mm for (a) friction tests and (b) impact tests.

The oxygen analyser has the following measuring range: 0 - 100% O_2. Additionally, an oxygen cylinder and an ignition source (voltage generator and spark plug) are necessary. The latter is employed to ignite the mixture only if the friction process has not produced the requisite hot sparks for ignition, thereby preventing the release of the mixture into the atmosphere. Metallic test specimens are composed of the pairs of materials to be tested, with the dimensions illustrated in Fig. 11(a) (dimensions are given in millimetres).

The impact test device, Fig. 12, uses a stand consisting mainly of: a table to provide sufficient energy for the test; the digital system for determining the fall height and calculating the impact energy; the test specimen; the steel plate; the blast chamber (the outer wall is open and covered with plastic film to allow the blast pressure to be released); the device for moving the plate longitudinally and transversely; the system for producing an explosive mixture, consisting of the dosing system and the oxygen analyser. The oxygen analyser has the following measuring range: 0-100% O_2.

Furthermore, an oxygen cylinder and an ignition source (comprising a voltage generator and a spark plug) are required. The ignition source is used to ensure that the mixture is ignited only if the impact process has not produced the hot sparks required, in order to prevent the mixture from being released into the atmosphere.

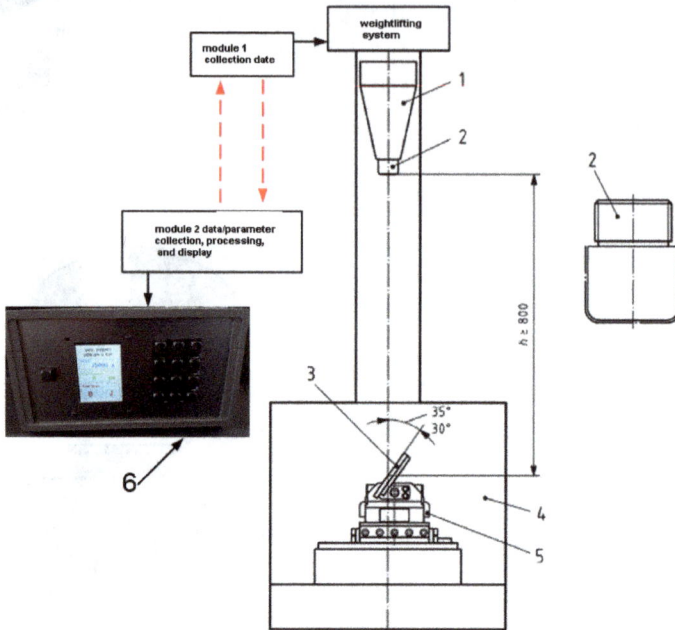

Figure 12. Assembly for impact ignition test, Legend: 1 - The mass with which the test energy is realised; 2 - The test specimen; 3 - A rusted steel plate; 4 - An explosion chamber (the outer wall of which is open and covered with plastic foil in order to allow the release of the explosion pressure); 5 - A device for longitudinal and transverse movement of the rusted plate; 6 - A digital system for determining the drop height and calculating the impact energy (module 1 + module 2), h - The drop height.

The findings of the impact test, with the test material sample exhibiting the configuration depicted in Fig. 11(b), (whether the test material passed or failed the test) are expressed by: firstly, whether or not ignition occurred in the initial 10 tests in the explosive mixture, or the number of ignitions in the subsequent 32 tests in the oxygen-enriched or double impact energy explosive mixture. The impact test is conducted within the laboratory and consists of the simulation on special stands of the spark formation process generated by accidental or technological impacts. The simulation is conducted on a dedicated test stand, where an impact is generated between a rusty plate positioned vertically and the specimen under evaluation. The determination of the impact energy is to be made using the digital fall height determination and impact energy calculation system. The formula used by the digital system to determine the impact energy is (1):

$$E = m \times g \times h \ [J] \tag{1}$$

Where m is the weight to be dropped in [kg], g is the acceleration of gravity in [m/s^2] and h is the height of the drop in [m]. The form of the weight is selected in accordance with the intended purpose of the material utilised in the construction of the equipment [23].

In order to avoid the production of hot sparks with the potential to ignite an explosive atmosphere, friction resistance experiments were conducted at the National Institute for Research and Development in Mine Safety and Explosion Protection (INSEMEX Petroşani) in accordance with the standard for impact and friction tests. The experimental materials were prepared and attached to steel plates (see Fig. 13 (a)), with a central area designated for wear in explosive atmospheres. The experiment was performed with steel wear elements (see Fig. 13(a)), and the contact, repeated 16,000 times, was between the steel element and the CuAlBe plates. For the impact tests, round steel balls were prepared and CuAlBe plates were fixed on the outer side (see Fig. 13(b)), to be exposed to impact with a steel plate.

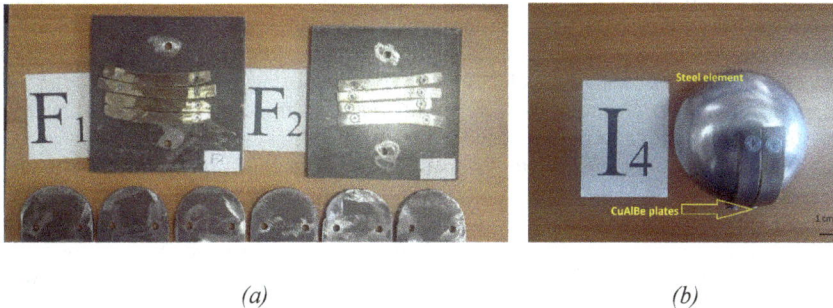

(a)   (b)

Figure 13. Items prepared for friction test in a) and impact test in b) in explosive atmosphere.

In this experiment, the impact test plates were subjected to thermal deformation directly onto the steel ball, and subsequently affixed mechanically to the surface of the ball. The exposed surface of the plates would come into contact with a substantial steel plate in an explosive atmosphere, thereby observing the formation of hot sparks and their capacity (or lack thereof) to ignite the explosive atmosphere.

In accordance with prevailing occupational health and safety legislation, the experiments were conducted in such a way as to eliminate all risks and hazards that could affect human resources during the experimental procedures [24].

2.5   Experimental methodology

The microstructural analysis was conducted using an optical microscope (Zeiss) equipped with a Moticam optical camera at a temperature of approximately 25°C. For the microstructural analysis, the samples were grinded with SiC paper to a grain size of 2400 and mechanically polished with an aqueous alumina suspension (1-0.3 μm). The chemical etching process was initiated using an aqueous solution of iron chloride (FeCl$_3$) for 10 seconds.

Scanning Electron Microscope (SEM) VegaTescan LMH II analyses were conducted for microstructural analysis, defect analysis of the obtained materials, analysis of the surface condition of the alloys after wear resistance tests, corrosion resistance tests and mechanical properties tests [25].

Non-destructive testing (NDT) was conducted following the melting of the alloys using fluorescent liquid penetrants. Subsequent to casting the experimental alloys and conducting electro-corrosion tests with a Bruker X-Flash EDS detector in Automatic, Element List, Mapping, Line and Point modes, the chemical composition was determined according to the type of investigation performed. The microstructure was determined after grinding up to 2000 with SiC paper, polishing with alumina solution and chemical etching with FeCl$_3$ for 2-5 seconds [26].

The present study utilised a VoltaLab 21 electrochemical system (PGP201-Economic Potentiostat), equipped with VoltaMaster 4 data acquisition and processing software, to analyse the electrochemical corrosion resistance. The corrosion analysis system consists of an electrochemical cell with three electrodes, a saturated calomel electrode as a reference electrode, a platinum electrode used as an auxiliary electrode and a working electrode consisting of the sample to be tested and a Teflon gripping system that exposes a certain area of the sample to the electrolyte. Prior to the commencement of the electro-chemical corrosion process, the sample was subjected to a sandblasting procedure, followed by a thorough cleaning and washing with distilled water. The electrolyte solution comprised NaCl 0.9% at room temperature. The linear polarization curves were obtained for a scan rate of the electrode potential of dV/dt=0.5 mV/s, and the cyclic polarization curves were obtained at a rate of 10 mV/s.

The electrochemical corrosion resistance of Cu-Ti alloys in the cast and rolled state was evaluated by linear and cyclic potentiometric tests using a VoltaLab 21 (PGP201-Economic Potentiostat) electrochemical system with a 3-electrode cell. The system was equipped with VoltaMaster 4 data acquisition and processing software. Prior to the commencement of the experiment, the samples were subjected to an electric current in a saline electrolyte medium and subsequently mechanically polished and washed with distilled water. A saline solution (NaCl 3.5%) at room temperature was used as the electrolyte solution. The linear polarization curves were plotted with the variation of the electrode potential rate dV/dt = 0.5 mV/s, and the cyclic one by polarization curves with a rate of 10 mV/s.

The polarization resistance method was employed to evaluate the corrosion rate. This method is based on determining the corrosion current at the corrosion potential of the metal or alloy from the linear polarisation curve obtained at relatively low overpotential. The corrosion current thus determined is that which occurs at the metal/corrosive medium interface when the metal is immersed in the solution, and it is considered to be the instantaneous corrosion current [27].

The corrosion current density was determined using Eg. 2:

$$J_{corr} = \frac{b_a \cdot b_c}{2.303(b_a + b_c) \cdot R_p} \quad (mA/cm^2) \tag{2}$$

Linear polarization curves were recorded at low potential rate (0.5 mV/s), obtained in the potential range ($E_{corr} \pm 150$ mV).

The investigation of the surface condition utilised optical microscopy, which employed a Zeiss microscope accompanied by a Moticam digital camera. In addition, this investigation involved scanning electron microscopy, VegaTescan LMH II SEM, SE detector and EDS spectroscopy, Bruker XFlash.

References

[1] J. Zhang, Y. Liu, G. Sha, S. Jin, Z. Hou, M. Bayat, N. Yang, Q. Tan, Y. Yin, S. Liu, J.H. Hattel, M. Dargusch, X. Huang, M. Zhang, Designing against phase and property heterogeneities in additively manufactured titanium alloys, Nat. Commun. 13 (2022) 4660. https://doi.org/10.1038/s41467-022-32446-2

[2] J. Wang, C. Liu, C. Leinenbach, U.E. Klotz, P.J. Uggowitzer, J.F. Löffler, Experimental investigation and thermodynamic assessment of the Cu-Sn-Ti ternary system, Calphad 35 (2011) 82-94. https://doi.org/10.1016/j.calphad.2010.12.006

[3] M.A. Turchanin, P.G. Agraval, A.N. Fesenko, A.R. Abdulov, Thermodynamics of liquid alloys and metastable phase transformations in the copper-titanium system, Powd. Metall. Met. Ceram. 44 (2005) 259-270. https://doi.org/10.1007/s11106-005-0090-6

[4] K.P. Kalinin, M.Z. Spiridonov, Studying the properties of copper-titanium alloys, Tr. Gos. Nauchn.-Issled. Proektn. Inst. Obrab. Tsvetn. Met. 18 (1960) 46-57.

[5] V.N. Eremenko, R.N. Mogilevskii, V.M. Sergeenkova, V.M. Petyukh, Melting heat and homogeneity range of TiCu intermetallic compound, Izv. Akad. Nauk. SSSR. Met. 6 (1988) 171-173.

[6] O. Taguchi, Y. Iijima, K. Hirano, Reaction diffusion in the Cu-Ti system, J. Jpn. Inst. Met. 54 (1990) 619-627. https://doi.org/10.2320/jinstmet1952.54.6_619

[7] T. Yamane, S. Nakajima, H. Araki, Y. Minamino, S. Saji, J. Takahashi, Y. Miyamoto, Partial phase diagrams of the titanium-rich region of the Ti-Cu system under high pressure, J. Mater. Sci. Lett. 13 (1994) 162-164. https://doi.org/10.1007/BF00278149

[8] S. Nagarjuna, D.S. Sarma, On the variation of lattice parameter of Cu solid solution

with solute content in Cu-Ti alloys, Scripta Mater. 41 (1999) 359-363.
https://doi.org/10.1016/S1359-6462(99)00187-6

[9] H. Yang, B. Zeng, H. Wang, H. Jin, C. Zhou, Phase equilibria and thermodynamic re-
assessment of the Cu-Ti system, Calphad 82 (2023) 102594.
https://doi.org/10.1016/j.calphad.2023.102594

[10] Y. Zhan, D. Peng, J. She, Phase equilibria of the Cu-Ti-Er system at 773 K and
stability of the CuTi₃ phase, Metall. Mater. Trans. A 43 (2012) 4015-4022.
https://doi.org/10.1007/s11661-012-1226-1

[11] A. Laik, K. Bhanumurthy, G.B. Kale, B.P. Kashyap, Diffusion characteristics in the
Cu-Ti system, Int. J. Mater. Res. 103 (2012) 661-672.
https://doi.org/10.3139/146.110685

[12] F. Sommer, K. Klappert, I. Arpshofen, B. Predel, Thermodynamic investigations of
liquid copper-titanium alloys, Z. Metallkd. 73 (1982) 581-584.
https://doi.org/10.1515/ijmr-1982-730909

[13] M. Arita, R. Kinaka, M. Someno, Application of the metal-hydrogen equilibration
for determining thermodynamic properties in the Ti-Cu system, Metall. Trans. A
10 (1979) 529-534. https://doi.org/10.1007/BF02658315

[14] C. Colinet, A. Pasturel, K.H.J. Buschow, Enthalpies of formation of Ti-Cu
intermetallic and amorphous phase, J. Alloys Compd. 247 (1997) 15-19.
https://doi.org/10.1016/S0925-8388(96)02590-X

[15] S.V. Konovalihin, I.I. Chuev, S.A. Guda, D.Yu. Kovalev, Estimation of enthalpy of
formation of TiCu by density-functional method, Phys. Met. Metallogr. 121 (2020)
1188-1292. https://doi.org/10.1134/S0031918X20120078

[16] K.C.H. Kumar, I. Ansara, P. Wollants, L. Delaey, Thermodynamic optimisation of
the Cu-Ti system, Z. Metallkd. 87 (1996) 666-672. https://doi.org/10.1515/ijmr-
1996-870811

[17] L.G. Zhang, L.B. Liu, G.X. Huang, Calphad 32 (2008) 527-534.
https://doi.org/10.1016/j.calphad.2008.05.002

[18] I.V. Nikolaenko, E.A. Beloborodova, G.I. Batalin, N.I. Frumina, V.S. Zhuravlev, Z.
Fiz. Khim. 57 (1983) 1154-1155.

[19] R.G. Chelariu, R. Cimpoesu, A.M. Jurca, C.M. Popa, M. Benchea, G. Badarau, B.
Istrate, A.M. Cazac, N. Cimpoesu, D.-D. Pintilie, Analysis of chemical,
microstructural and mechanical properties of a CuAlBe material regarding its role
as a non-sparking material, Materials 17 (2024) 2220.
https://doi.org/10.3390/ma17102220

[20] S. Bhaumik, V. Paleu, S. Datta, Tribological investigation of textured surfaces in
starved lubrication conditions, Materials 15 (2022) 1-10.
https://doi.org/10.3390/ma15238445

[21] N. Cimpoesu, V. Paleu, C. Panaghie, A.M. Roman, A.M. Cazac, L.I. Cioca, C.

Materials Research Foundations **186** (2026)   https://doi.org/10.21741/9781644903872

Bejinariu, S.C. Lupescu, M. Axinte, M. Popa, G. Zegan, Mechanical properties and wear resistance of biodegradable ZnMgY alloy, Metals 14 (2024) 1-15. https://doi.org/10.3390/met14070836

[22] SR EN ISO/IEC 17025:2018, Cerințe generale privind competența laboratoarelor de încercări și etalonări.

[23] R.G. Chelariu, C. Bejinariu, M.A. Bernevig, S.L. Toma, A.M. Cazac, N. Cimpoesu, Analysis of non-sparking metallic materials for potentially explosive atmospheres, in: Proceedings of the 10th International Conference on Manufacturing Science and Education - MSE 2021, Sibiu, Romania, 2021, pp. 1-6. https://doi.org/10.1051/matecconf/202134310014

[24] C. Bejinariu, D.C. Darabont, E.R. Baciu, I. Ionita, M.A.B. Sava, C. Baciu, Considerations on the method for self-assessment of safety at work, Environ. Eng. Manag. J. 16 (2017) 1395-1400. https://doi.org/10.30638/eemj.2017.151

[25] I. Hopulele, N. Cimpoeşu, C. Nejneru, Metode de analiză a materialelor. Microscopie şi analiză termică, Editura Tehnopress, Iaşi, 2009, ISBN 978-973-702-673-6.

[26] C. Munteanu, M. Ştefan, C. Baciu, N. Cimpoeşu, Metode difractometrice şi microscopie optică şi electronică în studiul materialelor, Editura Tehnopress, Iaşi, 2008, ISBN 978-973-702-563-0.

[27] E.E. Stansbury, R.A. Buchanan, Fundamentals of electrochemical corrosion, ASM Technical Books, 2000, ISBN 978-1-62708-302-7. https://doi.org/10.31399/asm.tb.fec.9781627083027

Materials Research Foundations **186** (2026)   https://doi.org/10.21741/9781644903872

CHAPTER 3

Experimental Results Obtained from the Analysis of CuAlBe Alloys

Romeo-Gabriel CHELARIU[1], Ramona Cimpoeşu[1]*,Gabriel-Dragoş VASILESCU[2],Costică Bejinariu[1,3]

[1]Faculty of Materials Science and Engineering, "Gheorghe Asachi" Technical University of Iasi, 67 Dimitrie Mangeron Street, 700050 Iasi, Romania

[2]National Institute for Research and Development in Mine Safety and Protection to Explosion— INSEMEX, 332047 Petrosani, Romania

[3]Academy of Romanian Scientists, Ilfov 3, 050044 Bucharest, Romania

ramona.cimpoesu@academic.tuiasi.ro

Abstract

For CuBe alloys used as non-sparking materials, an alternative with aluminium additions has been proposed to improve their properties and their application as driving elements such as gears. CuAlBe alloys have demonstrated remarkable performance in wear tests, where hot sparks are generated to ignite explosive atmospheres, and can be successfully used in anti-spark applications. The alloys obtained exhibit very good chemical and structural homogeneity, high machinability, and excellent deformability in hot-rolling deformation. The alloys demonstrate notable corrosion resistance in saline environments, forming a protective layer on the surface that mitigates the effects of surface corrosion. The mechanical properties of these alloys are excellent for a wide range of industrial applications, with small differences between the cast, hot-treated or rolled condition.

Keywords

CuAlBe Alloys, Anti-Spark, Wear Tests, Corrosion, Mechanical Properties

3.1 Chemical and structural analysis of experimental CuAlBe alloys obtained by conventional casting

The experimental alloys were obtained from high purity Cu (99.995%), CuBe alloy (Be 4%wt) and high purity aluminium (99.00%). In addition to the main elements identified by energy dispersive X-ray spectroscopy analysis, Table 1, trace amounts of Fe, Si and Mg were also identified whose total percentages did not exceed 0.2 wt%.

Table 1. Chemical composition of experimental CuAlBe alloys.

Chemical composition	Cu %		Al %		Be %		Other %
	wt	at	wt	at	wt	at	wt%
Alloy 1a	90.5	78	8.5	17.2	0.8	4.8	0.2
Alloy 2a	91.0	79.2	8.4	17.3	0.4	2.5	0.2

St. dev.: Cu: ±1.5; Al: ±0.5; Be: ±0.05.

From the literature review and starting from the known alloys CuBe (1-2% Be) for anti-spark applications and CuAl (approximately 10% Al) for various structural elements, the aim was to obtain a CuAlBe alloy to replace 1-2% Al with Be [1]. The charge was melted in an induction furnace in a ceramic crucible and remelted at 1100°C in a laboratory electric furnace to ensure chemical and structural homogeneity and to eliminate casting defects, details are shown in Fig. 1 with elemental mapping over an area of 1 mm². The average chemical composition of the alloys after 3 determinations on a 1 mm² area is given in Table 1.

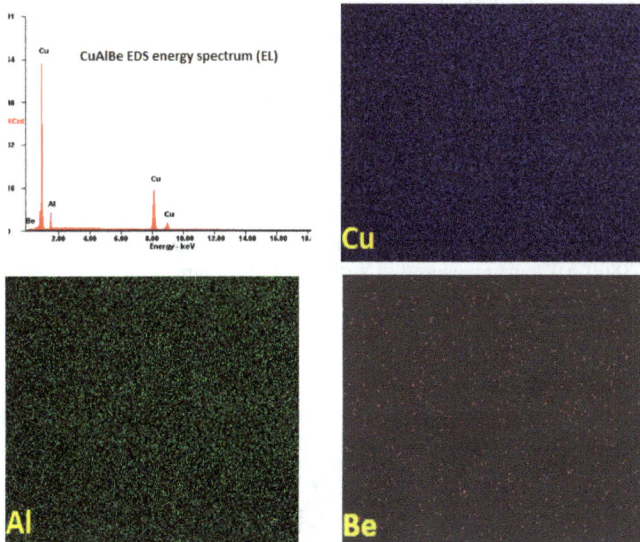

Figure 1. EDS spectra of CuAlBe sample and mapping of main elements (Cu, Al and Be).

The experimental alloys were subjected to heat treatment at 850 °C and water cooling, as well as hot rolling at 900 °C and rolling according to the scheme in Chapter 2, to produce ingots of 10 mm diameter and 100 mm length, and plates of 1.5 - 2 mm thickness and 180 mm length. Typically, depending on their chemical composition, bronzes are easily cold and hot rolled. In the case of CuAl alloys, it is recommended that the hot rolling process be performed in a heating range between 800 and 950 °C [2].

Beryllium bronzes are technical alloys containing 2-3% Be and having the phases: α - solid solution of substitution of Be in Cu, with CFC structure and γ - solid solution based on the CuBe compound, hard and brittle. Due to the variation of the solubility of Be in Cu with temperature, these bronzes can be hardened from 800°C to the α-structure, when they show a particular plasticity. Then, after an ageing heat treatment at 300-350°C, they can be precipitation hardened. Such bronzes are easily malleable, elastic, corrosion resistant, weldable and non-sparking. Applications include mining hammers and chisels, springs, diaphragms, watch parts, etc.

Beryllium is an alloying element that positively and significantly modifies the physico-mechanical properties, corrosion resistance, refractoriness and service life of aluminium bronzes. It also provides Al bronzes with anti-magnetic and anti-spark properties [3]. At high temperatures, these alloys exhibit a stable β-phase with a disordered structure which can be maintained at low temperatures by rapid cooling. The disordered β-phase is organised into a DO_3 structure during cooling at a slower solidification rate, decomposing into γ_2 and ά phases with lower or higher aluminium content. The martensitic transformation of the β-phase into 18R can be induced spontaneously by cooling or mechanically by stress.

The experimental alloys were cleaned in the ultrasonic chamber prior to NDT (non-destructive surface testing) analysis by a post-emulsification hydrophilic test. For this purpose, an ultra-high sensitivity level 4 penetrating solution and a hydrophilic emulsifier at a concentration of 5% were used. Dry powders were used for better contrast, which enhances pore localisation. The steps used were: penetrant holding time 20 min, emulsifier time 2 min and developer time 10 min. The parts were inspected under UV light with an intensity of 3800 $\mu W/cm^2$ measured at a distance of 38 cm from the lamp bulb [4]. After the melting/casting step, the materials were analysed on the surface using fluorescent penetrating liquids to observe if any voids, cracks or holes appeared in the alloy. Fig. 2 shows a clean surface for all three experimental alloys, indicating that a homogeneous material has been obtained.

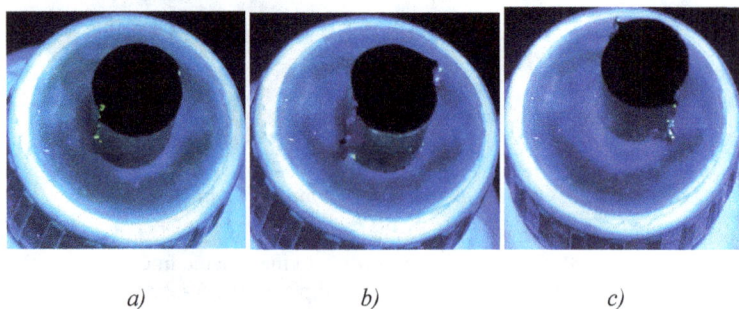

a)   *b)*   *c)*

Figure 2. NDT results a) Alloy 1, b) Alloy 2 and c) Alloy 3 after five re-melting operations.

By chemical composition analysis, the elements copper, aluminium and beryllium have been qualitatively identified in the energy spectrum in Fig. 3. A general analysis of the chemical element distribution, Fig. 3, shows a homogeneous chemical material with no specific conglomerations or depleted surface areas.

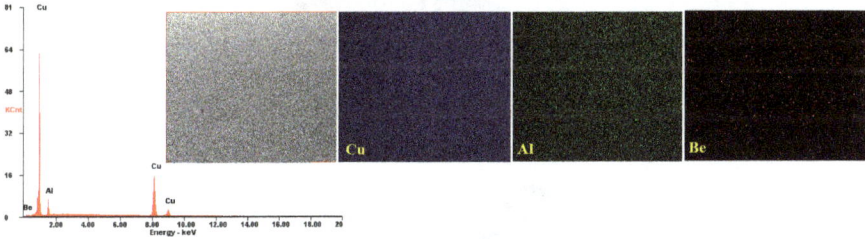

Figure 3. Determination of the chemical composition, energy spectrum and distribution of the main elements.

The quantitative chemical analysis, in mass % and atomic %, Table 2, shows the differences obtained in the chemical composition, labeled as experimental alloys: Alloy 1a, Alloy 2b and Alloy 3. The first two alloys considered for analysis have adequate percentages of Cu and Be with differences between Al and Alloy 2, while 3 has adequate percentages of Al and Cu and different mass percentages of Be. The influence of the additions (Al and Be) to the Cu matrix on the electro-corrosion resistance is investigated.

Table.2. Chemical composition (an average of 5 EDS detector determinations from different areas of the experimental materials)

Chemical	Alloy 1a		Alloy 2b		Alloy 3	
elements	wt%	at%	wt%	at%	wt%	at%
Cu	90.71	78.0	89.12	76.21	84.47	58.06
Al	8.5	17.22	10.41	20.97	10.32	16.7
Be	0.79	4.78	0.47	2.82	5.21	25.24

The microstructure of the alloys as seen by optical microscopy in Fig. 4 shows a large grain structure in all cases. The grains are large, hundreds of microns in all cases. The grain boundaries appear thick and show precipitates. In the case of Alloy 1a, Fig. 4 a), martensitic type plates can be observed. A homogeneous distribution of CuAl2 precipitates can be observed, especially in the first alloy and partially in the third [5]. The coarse compounds are located at the grain boundaries and at the shared grain boundaries, places that represent the lightest nucleation sites. Smaller grains show finer structural topography. Areas corresponding to precipitate boundaries show a lack of compounds, mainly on the Cu-Al matrix [6].

a)   *b)*

c)

Figure 4. Optical micrographs of the alloy structure (FeCl₃ etching) a) Alloy 1a, b) Alloy 2b and c) Alloy 3.

Differences in chemical composition due to the presence of precipitates between grain boundaries are an important and dominant factor in the electrocorrosion resistance of alloys when the potential difference between these two areas creates micropoles in the electrolyte solution. The structure of the third alloy, Fig. 4 c), shows a more homogeneous distribution of precipitates, not only at the grain boundaries but also within the grains [7].

Fig. 5 shows the sample microstructure and it can be observed that the predominant phase in all micrographs at room temperature is the austenite phase [8]. Structural evaluation by optical microscopy (OM) and scanning electron microscopy (SEM) of the microstructure reveals large grains typical of Cu-based alloys for all states, Fig. 5 a)-c) with an orientation in the rolling direction for the HR+WQ structure.

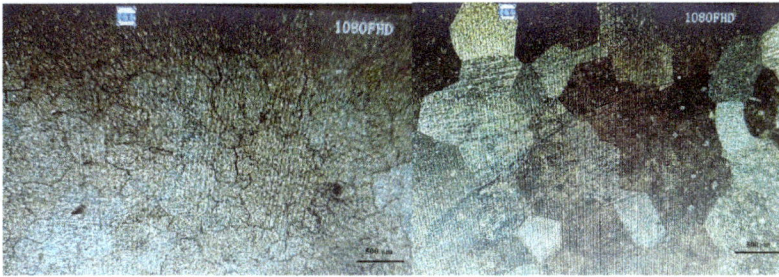

a)   *b)*

c)

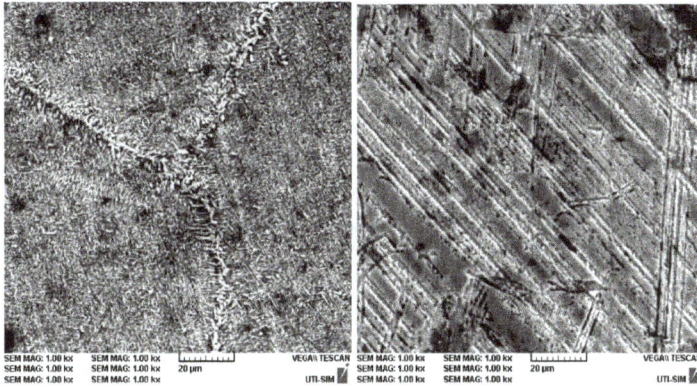

d)   *e)*

Figure 5. Optical (a)-c)) and scanning electron (d)-e)) microstructures of CuAlBe a) cast state of alloy 1, b) hot rolled state of alloy 1a, c) hot rolled + water quenched structure of alloy 1a, d) detail of alloy 1 after hot rolling and e) of alloy 2b after hot rolling and water quenching.

An abnormal grain growth (AGG) occurs after the hot rolling process compared to b) and c)) for both material states (HR and HR+WQ). This phenomenon has been reported by da Mota Candido et al. (da Mota Candido, 2020) in a previous study for a hot-rolled polycrystalline Cu-Al-Be alloy.

The induced plastic deformation of the material during hot rolling (passing the sample from an ingot of 8.5 mm diameter and 100 mm length through 6 passes into a plate of 2.5 mm thickness, 15 mm width and 140 mm length) is the reason for the dynamic recrystallisation process [9]. According to Oliveira et al., during the recrystallisation process by heat treatment + deformation, a whole new group of grains develops and a large energy contribution occurs during the deformation of the cast structure, e.g. grain intersections, sliding or self-locking of deformation lines, similar paths crossed by the sliding process and in some cases near grain boundary regions [10]. The abnormal grain growth caused by deformation in the hot rolling process is related to the grain boundary relocation process, which transforms (devours) the subgrains formed during the heating phase. In addition, the grain expansion rate increases with the degree (higher is higher) of small grain disorientation [9].

The main cause of AGG is the movement of grain boundary transforming subgrain boundaries that occurs during heating (850 °C) in each step of the hot rolling process, Chapter 2, schematically in Fig. 6(a) [11]. In this case of the plastic deformation of the CuAlBe alloy, there is a limit to the achievement of the deformation for the active stage of the recrystallisation step to occur and at this limit, when the deformation exceeds the critical value, the recrystallised grains are moderately restructured with the deformation [12]. The transition from austenite to martensite and vice versa is influenced by the low energy of straight boundaries and the absence of triple boundaries in polycrystalline materials [13].

3.2   Phase analysis of experimental alloys by X-ray diffraction (XRD)

In the cast state, the main phase is the martensitic phase, Fig. 6, consisting of the orthorhombic phase (18R- $\beta 1'$) and with reduced fractions of the monoclinic system (2H-$\gamma 1'$) at 2Θ angles of 33 and 47 [3]. By heat treatment, in addition to the appearance of the large and straight martensitic peak at the 2Θ angle of about 44, which is characteristic of the $\beta 1'$ (Cu_3Al) phase due to precipitation, a degradation of the martensitic transformation is also observed in the hot-rolled sample. This is mainly attributed to heating at 900 °C prior to plastic deformation by precipitation of the $\gamma 2$ (Cu_9Al_4) and α (lower Al content) phase by the emergence of the 2Θ peaks at 27 and 31, respectively.

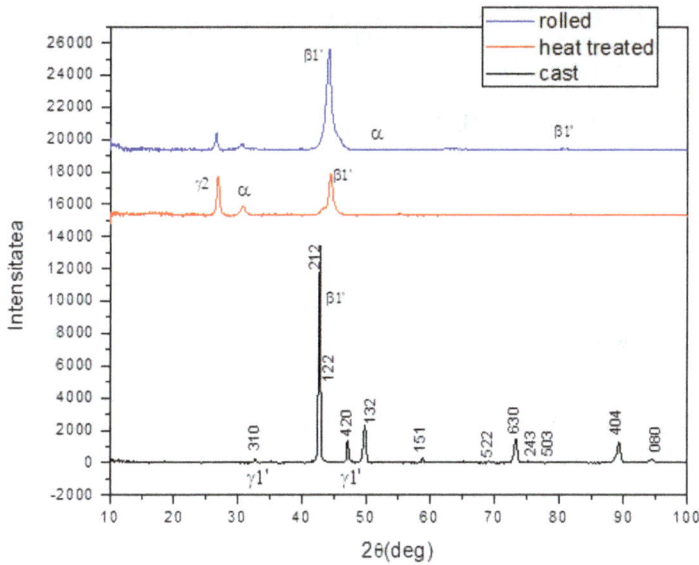

Figure 6. XRD patterns of CuAlBe alloy (Alloy 1a) in different states: cast, heat treated and hot rolled.

Fig. 6 shows the alloy after rolling in a single crystalline state with a low degree of polycrystallinity, reduced amorphous phase or defects. The single large peak that appears is the one with the (120) plane, indicating a martensitic β1' phase. To the right of this peak, another peak α (200) appears around the 2Θ of 50° and another martensitic β1' peak (042) around the 2Θ of 80° [14].

3.3   Electrochemical corrosion resistance of CuAlBe alloys analyzed

3.3.1   *Behavior of experimental CuAlBe-based alloys in a 3.5% NaCl saline solution*

Cu-based alloys generally exhibit high corrosion resistance and have been used in numerous applications in considerably corrosive environments. The addition of aluminum improves their corrosion resistance, especially in saline, chloride environments, as it helps to increase the protective nature of the composite layers [15,16]. Table 3 shows the experimental results for alloys 1a and 2b.

When corrosion occurs in an aqueous environment, copper dissolves anodically with the formation of ions [17]: Cu^{2+} și Cu^{+}:

$$Cu = Cu^{+} + e^{-}$$

$$Cu^{+} = Cu^{2+} + e^{-}$$

In the presence of HO^- and Cl^- ions, and depending on the pH of the solution, Cu^+ and Cu^{2+} ions can lead to insoluble products (Cu_2O, CuO, $Cu(OH)$, $Cu(OH)_2$) capable of forming protective barrier films by reactions such as:

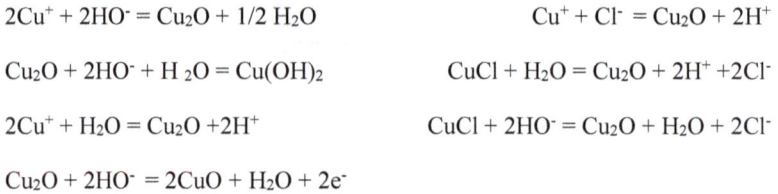

$$2Cu^+ + 2HO^- = Cu_2O + 1/2\ H_2O \qquad\qquad Cu^+ + Cl^- = Cu_2O + 2H^+$$

$$Cu_2O + 2HO^- + H_2O = Cu(OH)_2 \qquad\qquad CuCl + H_2O = Cu_2O + 2H^+ + 2Cl^-$$

$$2Cu^+ + H_2O = Cu_2O + 2H^+ \qquad\qquad CuCl + 2HO^- = Cu_2O + H_2O + 2Cl^-$$

$$Cu_2O + 2HO^- = 2CuO + H_2O + 2e^-$$

Literature studies indicate that corrosion occurs by dealumination, which favors the removal of aluminum-rich phases from the alloy [18]. The different behavior of CuAlBe alloys in chloride solutions is attributed to, first of all, their microstructure and their constituent phases. In samples with γ2 phase, a dissolution of precipitates occurs, protecting the matrix from the solid solution and reducing its desalinization process. On the other hand, precipitates with a higher Cu content, ά, have a higher corrosion stability. Linear potentiometry (Tafel curves in Fig. 7 a)) shows similar behavior of the 4 samples, grouped two by two according to their state: cast or hot-rolled and heat treated. The potential at which repassivation occurs is in the range of 180-200 mV/SCE for cast samples and 160-180 mV/SCE for rolled samples. The optical investigation of the alloys' surface showed the formation of a compound layer as a result of the interaction between the alloy and the electrolyte.

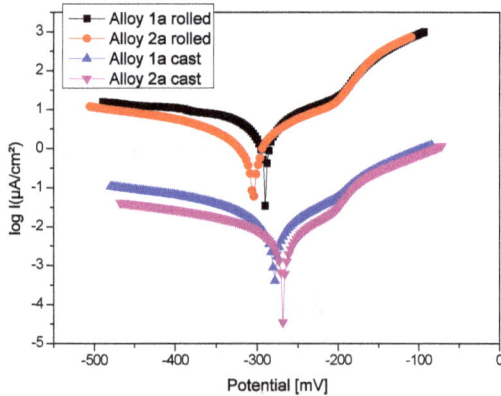

(a)

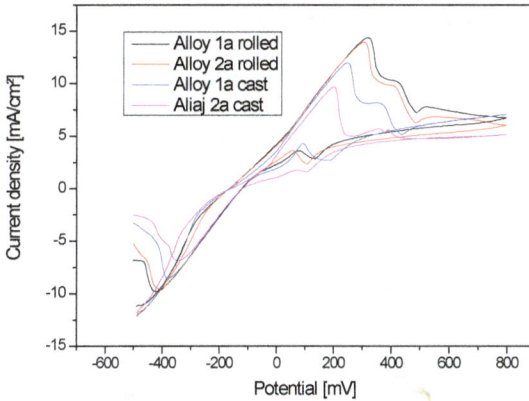

(b)

Figure 7. Results of linear and cyclic potentiometry. (a) Tafel diagram and (b) current density versus potential.

The main anodic reaction is the formation of copper and aluminum oxides and hydroxides, and the main cathodic reaction is oxygen reduction. Alumina can be formed on the surface by Al complexation with existing chlorides and a subsequent hydrolysis step. Table 3 shows the process parameters recorded during the electrochemical corrosion resistance test after 60 minutes of open-circuit potential analysis.

The cyclic potentiometry recorded and interpreted as a function of potential current density, Fig. 7 b), shows a pitting corrosion behavior of the alloy surface. The curves also show areas of surface passivation by the formation of copper oxides, which generally form a protective layer.

Table 3. Electrochemical corrosion test parameters.

Alloy	Linear potentiometry test parameters with Tafel interpretation					
	$-E(I=0)$ (mV)	i_{corr} (μA /cm$_c$	Rp kohm.cm$_c$	v_{corr} (μm/y)	$-\beta_c$ (mV/dec)	β_a (mV/dec)
Alloy 1a cast	278	31.6	1.86	436.0	378.0	122.2
Alloy 2a cast	268	30	3.77	414.5	725.2	123.8
Alloy 1a rolled	290	3.73	5.84	51.45	314.4	122.3
Alloy 2a rolled	306	3.49	14.53	48.11	379.6	83

Despite the fact that no variation in the corrosion resistance of CuAlBe alloys with the percentage of Be has been reported, a large difference in the behavior between CuAlBe samples with different phases in the composition has been observed. In this regard, the presence of other phases in addition to β will reduce the corrosion resistance of the alloy through the formation of local micro-piles between the phases. In the case of Cu-Al alloys, in addition to the dissolution and oxidation of Cu, the same corrosion process also occurs for Al. Thus, in addition to the Cl compound, Al salts such as AlCl and $Al(OH)_2Cl$ are incorporated on the surface.

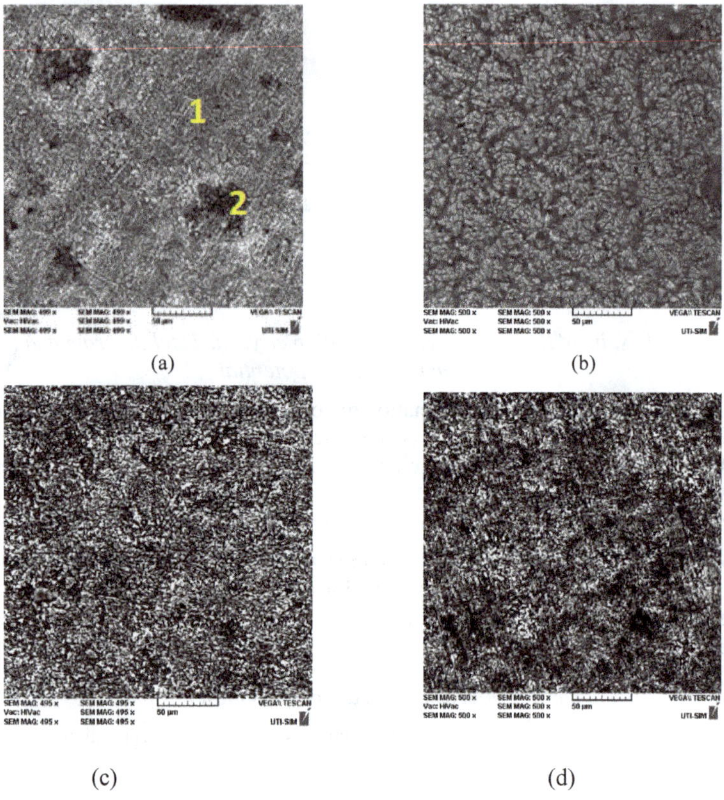

(a)   (b)

(c)   (d)

Figure 8. SEM images of CuAlBe surfaces after electrochemical corrosion tests (a) and (b) in the cast state and (c), (d) in the hot-rolled state.

It has been noted in the literature that one of the first corrosion products of Cu-based alloys in chloride environments is CuCl, which subsequently leads to the formation of Cu_2O copper oxide. Cu_2O is continuously oxidized to CuO for higher oxidation potentials [19]. On the surface of Cu alloys subjected to cyclic potentiometric tests, the presence of $CuCl_2$ compound was also identified. The alloy surfaces exhibited dealuminization, transforming the grain structure into a porous matrix rich in Cu. The detection of Al oxides on the surface corresponds to corrosion compounds formed in areas where the β-phase was dissolved. In all cases, Fig. 8 a) d), advanced surface corrosion is observed, due to the concentrated electrolyte solution used, of a zonal type, in particular of one phase. Oxide, carbide or chloride compounds can also be observed on the surface, Table 4 and Fig. 9.

The surface structure of alloy 1, Fig. 8 a) exhibits phases ($\beta + \gamma2$) after anodic polarization test in 3.5% NaCl [20], showing a pronounced surface depletion of Al, zone 1 and containing some corrosion products (zone 2: Al_2O_3, CuO, Cu_2O or even NaCl deposits).

On the matrix (zone 1), pits caused by severe dealuminization were observed, containing a porous product with a copper content of up to 87 wt%. In other areas of the sample, the Al_2O_3 film on the precipitate was not observed at all and a porous product rich in copper was also detected.

The presence of Al, O, Cl, Na or C based corrosion compounds on the alloy surface was confirmed by EDS analysis. The compounds that could form on the surface are: Al_2O_3, CuO, Cu_2O or even NaCl deposits. The presence of Be was also detected, indicating the formation of Be-based corrosion compounds. Open circuit potential values show that both alloys are prone to spontaneous corrosion when in contact with an electrolyte solution and especially when in direct ion exchange. Surface spectrometric analysis shows that the cast alloys are coated with a layer of more resistant compounds compared to the rolled alloys, due to a lower percentage of Cu identified on the surface.

Table 4. Chemical composition of cast and rolled alloys 1a and 2a.

Surface		Alloy 1 cast	Alloy 2 cast	Alloy 1 rolled	Alloy 2 rolled	EDS Error %
Cu%	wt	79.79	68.05	87.39	86.45	1.5
	at	53.24	34.45	56.68	55.64	
Al%	wt	0.55	0.89	0.32	0.17	0.1
	at	0.86	1.06	0.48	0.26	
Be%	wt	-	-	3.23	2.69	0.01
	at	-	-	14.75	12.22	
Na%	wt	9.02	4.88	1.23	2.01	3.93
	at	16.65	6.82	2.21	3.58	
C%	wt	6.13	9.82	6.91	7.51	10
	at	21.65	26.32	23.70	25.60	
Cl%	wt	2.98	1.38	0.16	0.18	1.2
	at	3.57	1.26	0.18	0.20	
O%	wt	1.54	14.98	0.77	0.97	0.5
	at	4.04	30.11	1.98	2.50	

St. Dev: Cu: ± 0.2; Al: ± 0.05; Be: ± 0.05; ± 2.1; Na: ± 0.6; C: ± 2; Cl: ± 0.25; O: ± 0.25

Materials Research Foundations **186** (2026)　　　　　　　　　https://doi.org/10.21741/9781644903872

As a result of the difficulties encountered in the detection of Be with most spectroscopic techniques, it was observed that in the case of the alloys formed, due to the state of the formed compounds on whose surface it is present, it could not be quantified, even if the distribution analysis showed that it was present on the surface, but in small amounts and areas. The elemental distributions on CuAlBe surfaces after corrosion tests are shown in Fig. 9 for the elements Cu, Al, Cl, Na and O only, because the element Be was not determined on the cast and corroded samples and the presence of C on the surface in Table 4 could be an error or misinterpretation of the EDS detector.

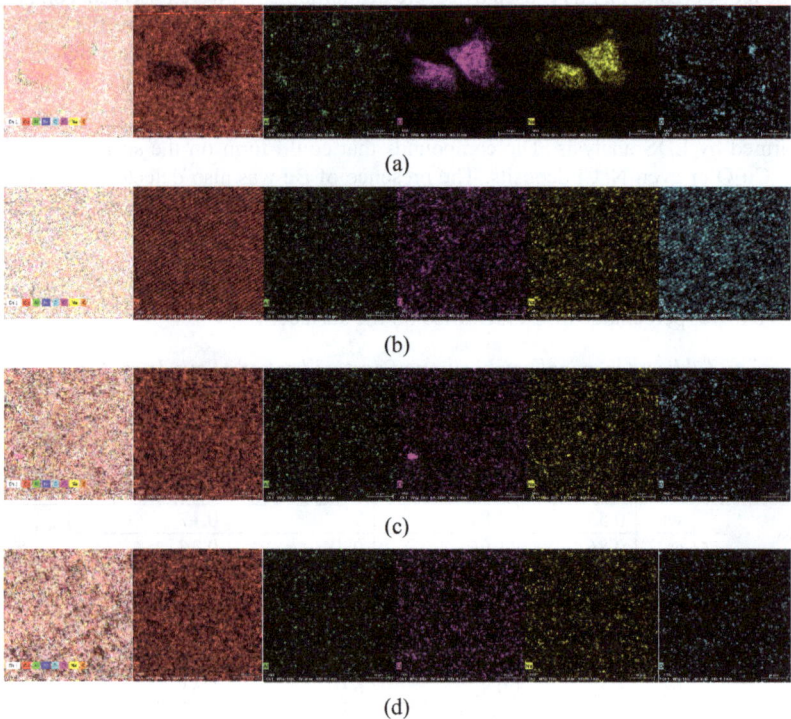

(a)

(b)

(c)

(d)

Figure 9. Elemental distribution on CuAlBe surfaces (Cu, Al, Cl, Na and O) after corrosion tests for (a) cast alloy 1a, (b) cast alloy 2a, (c) hot-rolled alloy 1a and (d) hot-rolled alloy 2a.

Fig. 9 shows the presence of salts, NaCl or oxides, mostly CuO-based, on the corroded surfaces. Most of the oxides have detached from the surface and migrated into the electrolyte solution.

3.3.2 *Electrochemical corrosion resistance in 0.9% NaCl saline solution*

Copper and its alloys are recognized as materials with good corrosion resistance and for this reason they are widely used in many fields of activity [21]. The good resistance of these materials is related to the formation on the surface of a uniform and adherent film (oxides, hydroxides, etc.) that protects the substrate from the environment.

The corrosion phenomenon is related to the formation and stability of this film; if the film is not formed or is destroyed, the metal is corroded, either generalized - over the whole surface - or localized (cracks, pitting). The results obtained for the three alloys are shown in Table 5.

Table 5. Electro-corrosion resistance test parameters.

Material/ parameters	OCP mV	E_0 mV	b_a mV	b_c mV	R_p kohm.cm²	J_{cor} µA/cm²	V_{cor} µm/y
Alloy 1a	-340	-342.9	168.5	-337.8	1.25	26.07	365.0
Alloy 2b	-304	-298.8	117.7	256.2	0.954	28.77	402.9
Alloy 3	-230	-260.8	112.1	103.5	1.45	10.83	151.6

Based on the data presented in Table 5, the following conclusions can be underlined: all three alloys have the open circuit potential and the corrosion potential with negative values, which emphasizes that, from a thermodynamic perspective, the natural tendency of these alloys is towards spontaneous corrosion. It is observed that increasing the percentage of Al [22], in the case of alloy 2 up to 10%, does not significantly affect the corrosion resistance. On the other hand, increasing the percentage of Be up to 5% in alloy 3 reduces the corrosion current density from 28.77 to 10.83 µA/cm².

A thorough examination of the cyclic corrosion diagram (see Fig. 10 b)) reveals the general characteristics of the corrosion process. It is observed that the current density does not vary with the potential. However, there were brief periods of passivation following the formation of Cu oxides. The third alloy exhibits a lower corrosion rate per year, less than half that of the first alloy. This alloy exhibits the highest percentage of Be and approximately ten percent aluminum [23].

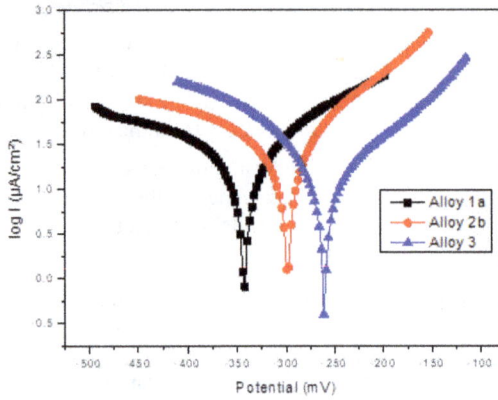

(a)

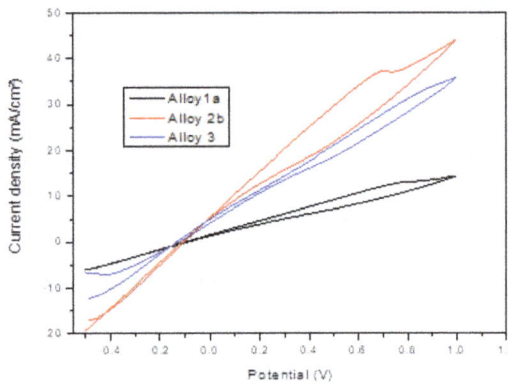

(b)

Figure 10. Electrochemical variations a) Tafel diagram and b) cyclic voltamograms.

Subsequent to the electrocorrosion tests in salt electrolyte solution, a rust layer was observed on the surface of the alloys. The chemical composition analysis revealed the presence of additional elements beyond the base elements: These include Cu, Al, and Be, along with a number of novel elements present in the salt electrolyte solution, particularly oxides and salts such as NaCl. These elements were identified through EDS analysis and their presence was observed in the energy spectrum in Fig. 11 a) [24]. In all cases, the surface was covered with a red oxide, copper oxide, with different morphologies (Fig. 11 b)-d)).

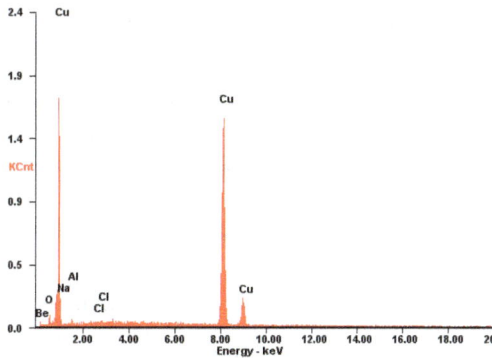

a)

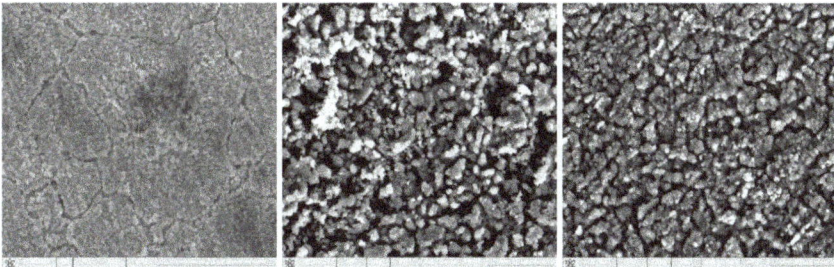

b)   c)   d)

*Figure 11. Alloy surface aspects after electrochemical corrosion test a) energy spectrum
b) alloy 1, c) alloy 2 and d) alloy 3.*

In the case of the initial alloy, the layer that covers the surface follows the positioning of the grains, with differences in boundaries. In contrast, the layer constituting the other two alloys, designated as 2 and 3, respectively, is comprised of numerous small oxides that manifest as round and rectangular shapes.

As illustrated in Table 6, the mean values of the chemical composition of the alloy surface were determined through five separate measurements in distinct regions. The electrocorrosion test method was utilized to observe the presence of oxygen, chlorine, and sodium. Typically, a composite oxide layer comprising CuO and Al_2O_3 is formed on the surface, providing a protective barrier against further corrosion [25]. The presence of the element Beryllium (Be) in the alloy has been observed to be higher in comparison to its initial state, a consequence of the Cu and Al oxidation. It is noteworthy that copper exhibits a more pronounced oxidation tendency when present with a third alloy.

Table 6. Surface chemical composition of alloys after electro-corrosion test.

Material/ Elements	Cu		Al		Be		O		Cl		Na	
	wt%	at%	wt%	at%	wt%	at%	wt%	at%	wt%	at%	wt%	at%
Alloy 1a	92.96	73.59	1.62	3.03	2.86	15.98	2.15	6.75	0.31	0.43	0.1	0.23
Alloy 2b	91.22	66.24	1.05	1.8	4.91	25.16	1.72	4.97	0.51	0.67	0.58	1.17
Alloy 3	85.29	52.65	1.71	2.49	7.72	33.59	3.32	8.13	0.33	0.37	1.63	2.79
EDS Error %	0.9		0.2		0.1		0.25		0.1		0.1	

The chlorine is derived from the NaCl salt, which migrates from the salt solution to the surface. The remaining chlorine is derived from Cl ions interacting with the metallic material. A notable observation is the significant decrease in aluminum percentage compared to the initial state, indicating a shift in the alloy composition. The predominant interaction of aluminum in the alloy is with oxygen, resulting in the formation of an alumina-like layer, with the majority of aluminum being present underneath or combined with copper oxide.

3.4   Mechanical properties of experimental CuAlBe alloys

3.4.1   *Indentation analysis of experimental alloy characteristics*

The modulus of elasticity (E) was estimated from load-displacement curves (Fig. 12) using the Oliver and Pharr method. The mechanical properties showed variations between the hot rolled and water quenched states based on the partial transformation from austenite to martensite. Furthermore, the mechanical properties varied compared to the cast state due to the relationship between abnormal grain growth phenomena and the elimination of typical casting defects.

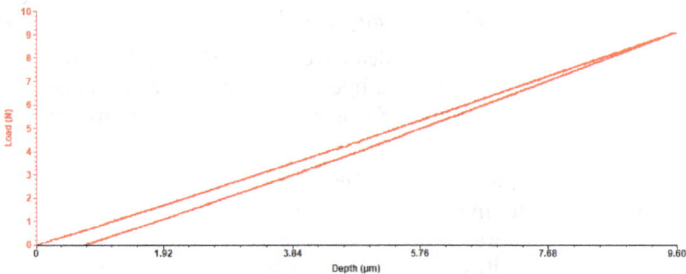

a)

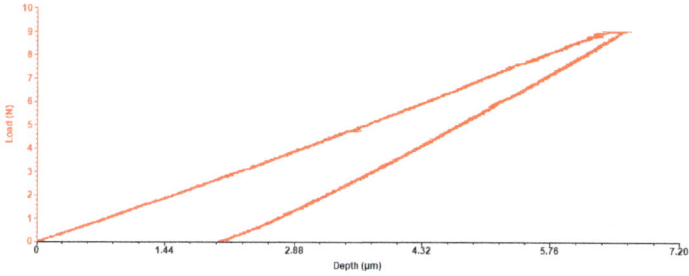

b)

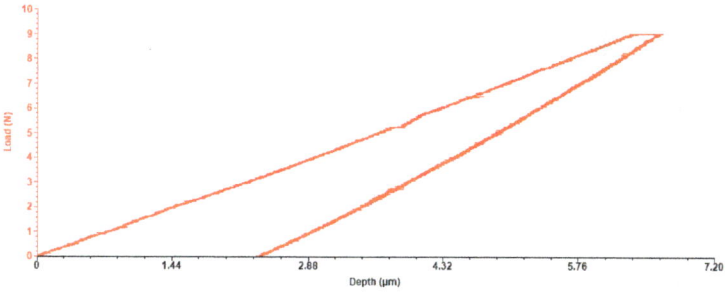

c)

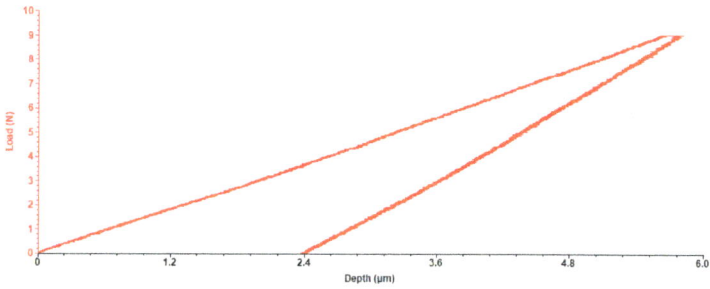

d)

Figure 12. Indentation plots for CuAlBe samples (1a and 2a) a) hot-rolled sample 1a, b) hot-rolled sample 2a, c) heat-treated (water quenched) hot-rolled sample 1 and d) heat-treated (water quenched) hot-rolled sample 2a.

As shown in Fig. 12, the mechanical properties of the materials are listed in Table 6. In order to establish the homogeneity of the properties throughout the material, five determinations were performed in different areas (points 1-5).

Table 6. Mechanical properties of experimental CuAlBe alloys in the hot-rolled and heat-treated state obtained from indentation tests [26].

Material	Area	Young indentation modulus [GPa]	Hardness [GPa]	Contact stiffness [N/μm]	Maximum load [N]	Maximum displacement [μm]	Contact depth [μm]	Contact area [μm²]
CuAlBe_WQ 1a	Area 1	13.96	2.13	1.1	9.05	9.58	3.41	4248.61
	Area 2	14.21	2.16	1.11	9.03	9.46	3.36	4183.68
	Area 3	14.06	2.15	1.10	9.03	9.52	3.38	4207.26
	Area 4	14.06	2.15	1.10	8.98	9.49	3.35	4171.78
	Area 5	13.82	2.11	1.09	9.02	9.63	3.44	4279.40
	Average	**14.02**	**2.14**	**1.10**	**9.02**	**9.53**	**3.38**	**4218.15**
CuAlBe_WQ 2a	Area 1	27.75	2.01	2.22	9.02	6.66	3.61	4495.87
	Area 2	25.42	2.16	1.97	9.02	6.79	3.38	4183.62
	Area 3	23.47	2.21	179	9.03	7.05	3.27	4079.49
	Area 4	20.58	2.18	1.58	8.99	7.56	3.30	4112.85
	Area 5	17.41	2.16	1.41	9.02	8.43	3.65	4539.79
	Average	**22.92**	**2.14**	**1.79**	**9.02**	**7.29**	**3.44**	**4282.32**
CuAlBe_HR 1a	Area 1	30.65	1.90.	2.38	9.02	6.65	3.81	4743.19
	Area 2	30.12	2.15	2.33	9.03	6.29	3.38	4205.28
	Area 3	29.60	2.12	2.30	9.04	6.36	3.42	4260.12
	Area 4	28.49	2.05	2.26	9.04	6.55	3.55	4416.16
	Area 5	28.40	2.14	2.20	9.01	6.45	3.37	4202.27
	Average	**29.450**	**2.07**	**2.29**	**9.024**	**6.46**	**3.50**	**4365.40**
CuAlBe_HR 2a	Area 1	37.16	2.09	2.89	9.03	5.81	3.47	4321.45
	Area 2	37.44	2.29	2.78	9.04	5.60	3.16	3934.70
	Area 3	40.1	2.341	2.95	9.01	5.380	3.086	3847.94
	Area 4	40.22	2.48	2.87	9.05	5.29	2.93	3652.13
	Area 5	42.16	2.48	2.87	9.02	5.27	2.911	3631.16
	Average	**39.42**	**2.34**	**2.87**	**9.03**	**5.47**	**3.11**	**3877.47**

The differences between the HR and WQ states regarding the mechanical properties, in particular the Young's modulus at indentation (GPa), are due to the presence of a more martensitic state in the WQ samples (the Young's modulus of A-phase austenite and M-phase martensitic for a CuAlBe alloy is known to be 75,000 MPa and 35,000 MPa, respectively [27]. In comparison with the molten state, an increase in Young's modulus (values for the cast state are published in another paper and are Young's modulus [GPa]: 23.9 and 17.2 respectively) is observed for both states (especially for hot rolled and less for water quenched samples) and also an increase in hardness due to the hot rolling process. Conversely, the hardness of the second alloy demonstrates a decline upon heat treatment (a higher percentage of the martensitic phase being transformed by water quenching). The contact stiffness also shows an important change due to the phase transformation from austenite to martensite, based on the known fact that the martensitic phase is softer and the austenitic phase is stiffer.

3.4.2   Scratch test results

From a mechanical behaviour perspective, Cu, CuBe and the majority of copper-based alloys are classified as elasto-plastic materials. At low levels of stress, the samples exhibit a linear elastic zone, and for loads higher than the yield stress, they undergo irreversible plastic deformation with broad deformation limits. The experimental results of the scratch strength tests conducted on the two specimens are shown in Fig. 13 and Table 7, with no significant variation depending on the Be content. The recorded variations included Fx (material resistance force to scratch force), CoF (coefficient of friction) and AE (acoustic emission: the occurrence of acoustic (elastic) undulation propagation in solid materials) over a distance of 10 mm.

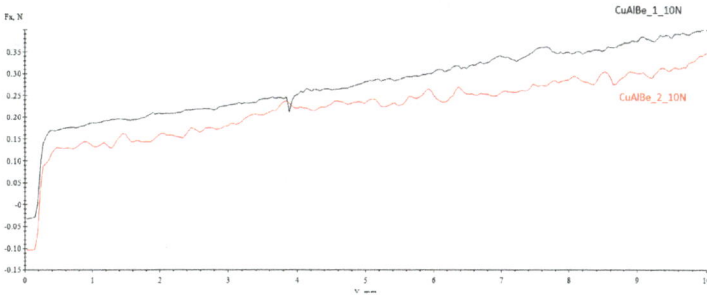

(a)

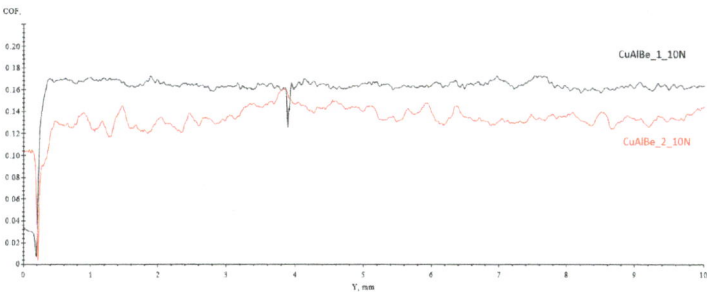

(b)

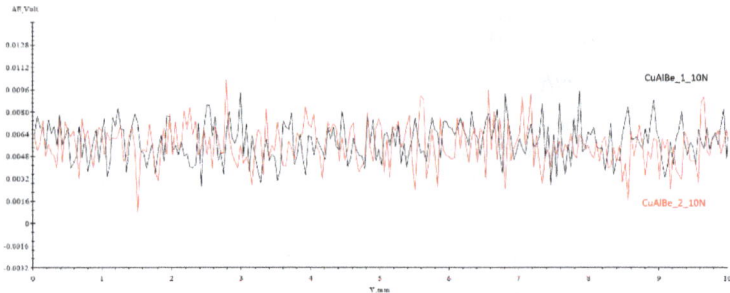

(c)

Figure 13. Variations of the surface reaction force of the cast alloy in (a) coefficient of friction (CoF) in (b) and acoustic emission (AE) in (c).

CoF exhibits minor surface area variations, which may be caused by grain boundaries, stress-induced martensite zones or different metal compounds with a lower coefficient of friction compared to the copper-based solid solution [28]. Stress-induced martensite (M) or residual martensite areas are softer in terms of hardness and stiffness compared to the rest of the material, which is in an austenitic state. The coefficient of friction for alloy 1 (Fig. 14(b)) is slightly higher due to the lower percentage of Be in the alloy and the lower microhardness due to the very low number of intermetallic compounds with Be.

Table 4.7. Mechanical properties of cast alloys (average).

Material	F_x [N]	AE [V]	COF [-]	F_f [N]	Fz [N]
CuAlBe_1a_10N	0.043	0.0058	0.179	1.043	5.24
CuAlBe_2a_10N	1.71	0.006	0.274	1.713	5.47

As shown by the variations in acoustic emission, Fig. 14 c), no effects of crack propagation or plastic deformation of the material due to ageing, temperature gradients or external stresses were observed [29]. The values determined from the scratch test, as presented in Table 7, indicate a higher coefficient of friction for sample 2, the alloy with a higher mass percentage of Be. This outcome indicates that the presence of Be in a higher percentage favours the formation of stress-induced martensite, a phase that is softer than β austenite, which leads to an increase in the coefficient of friction of the alloy. The acoustic emission values are found to be highly similar, and the shape of the variation curves confirms the polycrystallinity of the alloys and the existence of different phases and/or compounds with different stiffness [30].

The scratch test on heat-treated and rolled samples reveals the frictional properties of the materials. As shown in Fig. 14 for sample 1, the variations in Fx (material resistance to scratch force), COF (coefficient of friction) and AE (acoustic emission) represent the occurrence of acoustic (elastic) wave propagation in solid materials. This phenomenon occurs when a material exhibits irreversible changes in its microstructure, especially at the surface and near the interior of the material, as a consequence of crack growth and development or as a plastic deformation of the material due to aging, temperature gradients or external stresses [31] over a distance of 10 mm. On the heat-treated alloy, a difference in the Fx force can be observed between 4 and 7 mm after the test, when the tip of the equipment encounters a different phase, the martensite zone, after which it returns to the initial behavior.

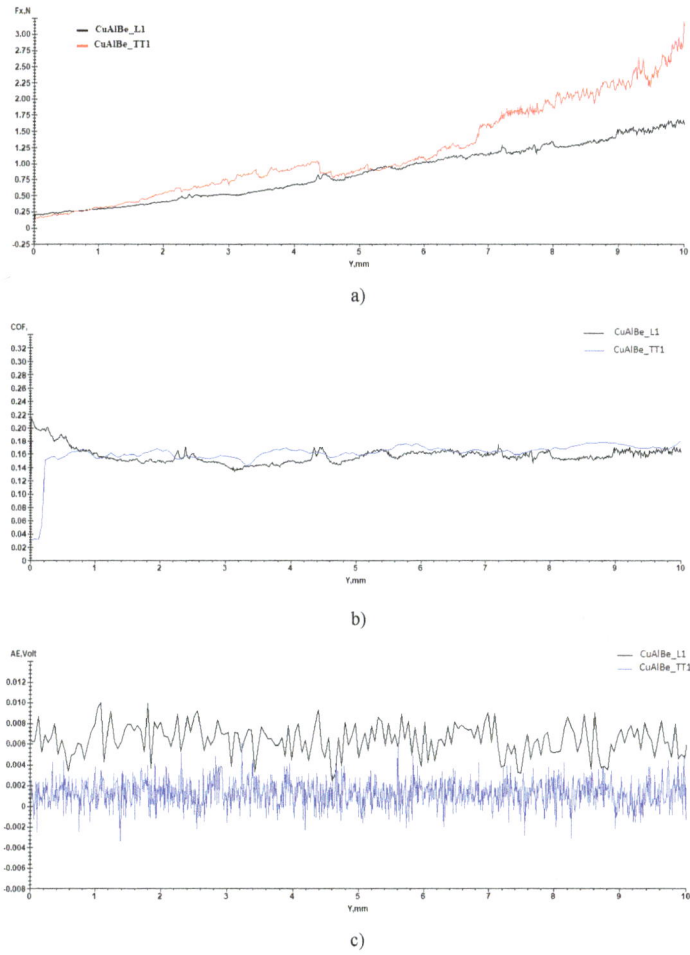

a)

b)

c)

Figure 14. Variations of scratch test results (a) Fx, (b) COF and (c) AE over a length of 10 mm for rolled and heat treated alloy 1a.

59

The behaviour of the coefficient of friction is similar for both samples (states), Fig. 13 b), with different increases that may correspond to grain boundaries where different metal compounds with higher COF are outlined (Chelariu, 2023). The martensite state (M) of the materials, characterised by the transformation from austenite to martensite, is a softer state in comparison to the stiffer austenite. This influence on the coefficient of friction is shown, for example, in the case of sample 1, the water quenching heat treatment favours the transformation from austenite to martensite and a higher percentage of M contributes to the increase of the COF value, Table 8.

Table 8 Mechanical properties of experimental CuAlBe alloys in the hot-rolled and heat-treated state obtained from scratch tests.

Material	Fx [N] Average	Ae [Volt] Average	COF [-] Average	Ff [N] Average
CuAlBe_HR1a	0.85	0.005	0.16	0.85
CuAlBe_HR2a	0.94	0.005	0.17	0.94
CuAlBe_WQ1a	1.16	0.001	0.21	1.16
CuAlBe_WQ2a	0. 98	0.001	0.18	0.97

In comparison with the cast condition, a change in the coefficient of friction is observed only for the second alloy (a decrease from 0.27 to 0.17) and no effect of hot rolling on the COF for the first sample. The average values of the coefficients of friction demonstrate an increase with heat treatment in the case of the first alloy based on the occurrence of the martensite phase. Acoustic emission analysis reveals similar values for hot rolled and water quenched samples.

3.5   Optical surface evaluation of CuAlBE samples after the explosive severe wear test

The necessity for explosion protection is increasing as explosions threaten the life and integrity of workers due to the accidental consequences of fire and pressure, the possibility of hazardous process compounds and the use of O_2 from the ambient atmosphere breathed by workers. The occurrence of a deflagration is initiated when a gaseous fuel, mixed with air (a minimal amount of O_2), reaches the vicinity of an electric furnace in the presence of an ignition source.

The use of robust materials and proactive construction is recommended in order to prevent the ignition of flammable substances in hazardous locations. In contrast to ordinary work or electronic equipment, this state-of-the-art equipment is designed to reduce sparks through the use of special non-ferrous materials [32]. Explosion-proof equipment with non-sparking components (more on this later) can be found in the OSHA 1910 guidelines. 146 Appendix D: A pre-entry checklist suggests that power tools used in enclosed spaces should not produce sparks [33]. Electrical sparks are a primary ignition source in many instances of equipment and tools around industrial facilities. In order to reduce the occurrence of spark generation, explosion-resistant equipment is typically constructed from non-sparking materials. In accordance with this principle, the majority of non-ferrous metals employed in the fabrication of explosion-proof systems exhibit elevated thermal conductivity [34]. In addition, non-metallic, non-sparking materials are frequently employed in explosion-proof equipment (Chelariu, 2023c). Examples of such materials include plastics, wood, thermoplastic polymers and leather. These materials are primarily employed in the fabrication of non-sparking tools, such as hammers, shovels and gloves [35]. In the context of explosion-proof lighting systems, robust non-sparking polymers such as polycarbonate or carbon fibre are employed to prevent the generation of sparks. This sub-chapter presents the initial findings from experiments conducted on CuAlBe alloys, with two different chemical compositions and in a rolled state, as non-sparking materials through standardized wear tests.

It is considered that Be bronzes are intended for the manufacture of hard, non-sparking tools (hammers, chisels, pliers, etc.) used in refineries, chemical industries and mining, special bearings, wear parts, gears, cams, stainless spiral springs, wear springs, parts for antimagnetic clocks, bellows and diaphragms in measuring instruments and automation, pressure welding electrodes (which also contain Co), electrical contacts and arc blades in electrical circuits. The authors propose the solution of a Cu-Al-Be alloy for the manufacture of elements, such as gears, which operate in an explosive atmosphere and also perform mechanical work during operation. As demonstrated by the experimental tests (severe wear following sparking during contact with metals), the surface of the plates exhibited a removal of the oxide layer (Fig. 15 (a) and (b)). The most affected plates were the centre plates, where the action of the steel counterpart was more intense (Fig. 15 (b)).

No wear spots were observed on the steel support of the plates, i.e. the entire contact area between the CuAlBe plates and the steel wear elements, Fig.15 (a) and (b). In the case of the centrally placed plates (see Fig. 15 (b)), large cracks can be observed on the wear area. The softer material, CuAlBe, is observed on the corresponding steel elements, Fig. 15 (c) and (d), confirming the severe aspect of the wear test by removing the material from the plates. It is therefore concluded that the presence of CuAlBe elements in explosive atmospheres is incompatible with the hardness required for the function of the elements. As illustrated in Fig. 16, the image provides an optical representation of the worn area. Material overlaps are observed due to the wear of hard materials. No cracks or pores were identified in this area.

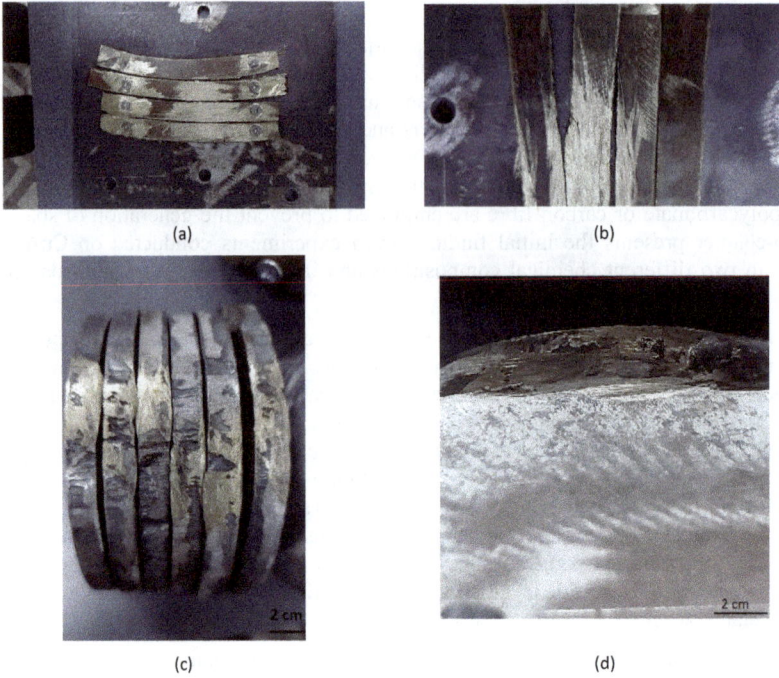

(a)

(b)

(c)

(d)

Figure 15. Worn plates after wear resistance test, (a) first set of plates - CuAlBe alloy 1a, (b) second - CuAlBe alloy 1b set of CuAlBe plates, (c) and (d) steel elements used for forced wear of CuAlBe plates.

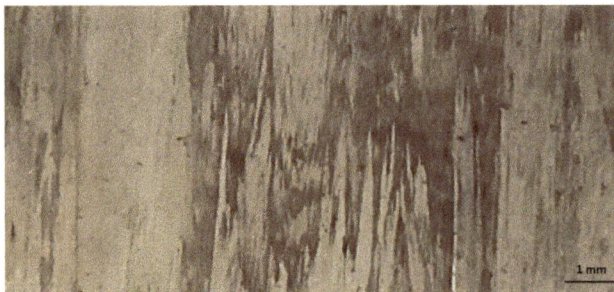

Fig. 16 Optical image of the worn surface.

The specialists from the National Institute for Research and Development in Mine Safety and Explosion Protection (INSEMEX Petroșani) have summarized the main conclusions of the explosive atmosphere wear tests in Table 9. CuAlBe alloys can exhibit pseudoelastic behavior at room temperature [36]. The hysteresis curve is associated with the dissipation of mechanical energy into heat. These properties enable these alloys to be used as external stress damping elements. The non-sparking character exhibited by these alloys is attributed to their high capacity to absorb mechanical stresses, thereby mitigating the formation of hot sparks.

The sparks resulting from the friction process are considered harmless (the tested material pair passes or fails the test) if:

- No ignition occurs during the first 16,000 frictions in the explosive mixture;

- No more than 8 ignitions occur during the subsequent 16,000 frictions in the explosive mixture enriched with up to 25% oxygen;

Table 4.9. Results after friction test in explosive atmosphere.

Nr. crt.	Sample type	Explosive mixture (10% H_2) or 6.5% CH_4	Material pair	Ignition during the first 16 000 friction cycles in explosive mixture (YES / NO)	Ignition during the first 16 000 friction cycles in explosive mixture enriched with O_2 up to 25% (YES/ NO)
1	Samples CuAlBe_1	10% H_2	F1-CuAlBe_1 and steel	NO	NO
2	Samples CuAlBe_2	10%H_2	F1-CuAlBe_2 and steel	NO	NO

In conclusion, considering the aforementioned test results and the test conditions, the tested materials (F1 samples) are considered non-sparking materials.

The contact intensity is observed as the CuAlBe material passes over the steel wear plates, as can be seen in the middle of Fig. 15 (c). In addition to the passage of the lower hardness material, Al-bronze, on the steel used for friction, several cracks can be observed on the material plates in the right of Fig. 15 [37]. These cracks are attributed to alloy group 1, in which the austenite phase predominates. Consequently, the plates of alloy group 2 exhibited similar behaviour in terms of generating burst-triggering sparks and were used. The samples that exhibited severe wear and were in the rolled and heat-treated condition did not produce burst-triggering sparks, regardless of the component phases, one or more [3].

Fig. 17 shows the surface condition of the experimental alloys after wear. Evidence of material detachment and the subsequent formation of corrosion oxides is observed on the surface.

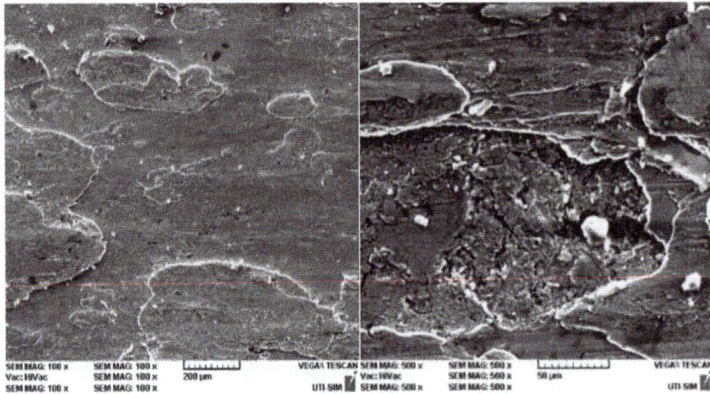

(a)   (b)

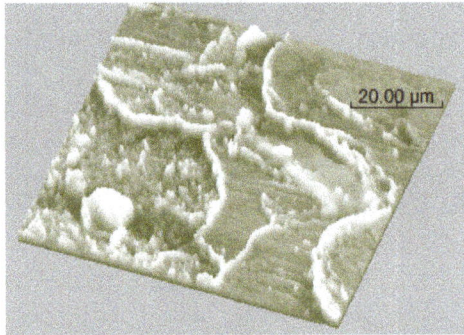

(c)

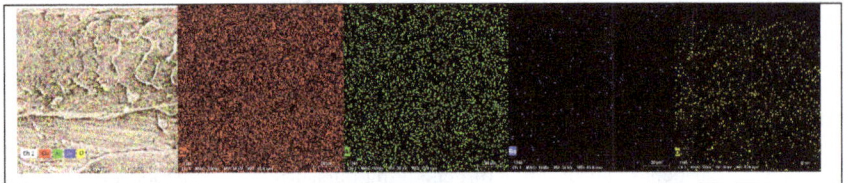

(d)

Figure 17. SEM images of the surface after wear test (a)500x, (b) 1000x, (c) 1500x - 3D and (d) elemental distributions on the worn area.

Sparks resulting from the friction process are considered non-hazardous (the pair of tested materials passed or failed the test) if: no ignition occurs during the first 16,000 friction cycles in the explosive mixture, or no more than eight ignitions occur during the next 16,000 friction cycles in the explosive mixture enriched with oxygen up to 25%. It can be observed that the alloy has been exfoliated from the surface, Fig. 17 c), in several layers for different stresses depending on the number of passes made. Surface oxidation, as Fig. 17 d) shows, generally does not have a significant effect on the generation of material sparking.

References

[1] N. Venugopal, V. Ramanathan, Experimental investigation on performance of aluminum alloy 7068 gears in a motorcycle gearbox, Int. J. Mech. Prod. Eng. 5 (2017) 1-6.

[2] L.D. Huang, C. Yan, Y. Zhou, W. Yan, Multi-directional forging and aging treatment effects on friction and wear characterization of aluminium-bronze alloy, Mater. Charact. 167 (2020) 110511. https://doi.org/10.1016/j.matchar.2020.110511

[3] R.G. Chelariu, R. Cimpoesu, A.M. Jurca, C.M. Popa, M. Benchea, G. Badarau, B. Istrate, A.M. Cazac, N. Cimpoesu, D.-D. Pintilie, G.D. Vasilescu, C. Bejinariu, Analysis of chemical, microstructural and mechanical properties of a CuAlBe material regarding its role as a non-sparking material, Materials 17 (2024) 2220. https://doi.org/10.3390/ma17102220

[4] C. Panaghie, R. Cimpoesu, B. Istrate, N. Cimpoesu, M.A. Bernevig, G. Zegan, A.M. Roman, R. Chelariu, A. Sodor, New Zn3Mg-xy alloys: characteristics, microstructural evolution and corrosion behavior, Materials 14 (2021) 2505. https://doi.org/10.3390/ma14102505

[5] S. Montecinos, A. Cuniberti, Effects of grain size on plastic deformation in a β CuAlBe shape memory alloy, Mater. Sci. Eng. A 600 (2014) 176-180. https://doi.org/10.1016/j.msea.2014.02.028

[6] Z. Song, S. Kishimoto, J. Zhu, Y. Wang, Study of stabilization of CuAlBe alloy during martensitic transformation by internal friction, Solid State Commun. 139 (2006) 235-239. https://doi.org/10.1016/j.ssc.2006.05.037

[7] R.G. Chelariu, N. Cimpoesu, T.I. Birnoveanu, B. Istrate, C. Baciu, C. Bejinariu, Obtaining and analysis of a new aluminium bronze material using induction furnace, Arch. Metall. Mater. 67 (2022) 1251-1257. https://doi.org/10.24425/amm.2022.141049

[8] S. Montecinos, A. Cuniberti, Thermomechanical behavior of a CuAlBe shape memory alloy, J. Alloys Compd. 457 (2008) 332-336. https://doi.org/10.1016/j.jallcom.2007.03.077

[9] S. Biswas, P.C. Gautam, A.J. Shukla, D.K. Chouhan, Dynamic recrystallization and its effect on microstructure and texture evolution in magnesium alloys, Encycl. Smart Mater. (2021) 476-481. https://doi.org/10.1016/B978-0-12-815732-9.00016-4

[10] S. Montecinos, S.N. Simison, Study of the corrosion products formed on a multiphase CuAlBe alloy in a sodium chloride solution by micro-Raman and in

situ AFM measurements, Appl. Surf. Sci. 257 (2011) 7732-7738.
https://doi.org/10.1016/j.apsusc.2011.04.018

[11] R.G. Chelariu, N. Cimpoeșu, A.M. Cazac, M.A. Bernevig, C. Bejinariu, Optical
evaluation of CuAlBe samples after harsh wear tests in explosive atmosphere, in:
Proc. 11th Int. Symp. Occupational Health and Safety (SESAM 2023), Bucharest,
2023, pp. 1-6.

[12] Y.R. Su, T.H. Wu, I.C. Cheng, Synthesis and catalytical properties of hierarchical
nanoporous copper from θ and η phases in CuAl alloys, J. Phys. Chem. Solids 151
(2021) 109915. https://doi.org/10.1016/j.jpcs.2020.109915

[13] V.H.C. de Albuquerque, T.A.A. de Melo, R.M. Gomes, S.J.G. de Lima, J.M.R.S.
Tavares, Grain size and temperature influence on the toughness of a CuAlBe shape
memory alloy, Mater. Sci. Eng. A 528 (2010) 459-466.
https://doi.org/10.1016/j.msea.2010.09.034

[14] Y. Li, Y. Lian, Y. Sun, Comparison of cavitation erosion behaviors between as-cast
and friction stir processed Ni-Al bronze in distilled water and artificial seawater, J.
Mater. Res. Technol. 13 (2021) 906-918.
https://doi.org/10.1016/j.jmrt.2021.05.015

[15] T. Ma, B. Tan, L.G. Savaş, K.Z. Kao, S. Zhang, R. Wang, N. Zeng, Y. He,
Multidimensional insights into corrosion inhibition of potassium oleate on Cu in
alkaline medium, Mater. Sci. Eng. B 272 (2021) 115330.
https://doi.org/10.1016/j.mseb.2021.115330

[16] C. Zeng, B. Zhan, A.H. Ettefagh, H. Wen, H. Yao, W.J. Meng, S. Guo, Mechanical,
thermal and corrosion properties of Cu-10Sn alloy prepared by laser powder bed
fusion, Addit. Manuf. 35 (2020) 101411.
https://doi.org/10.1016/j.addma.2020.101411

[17] G. Kear, B.D. Barker, F.C. Walsh, Electrochemical corrosion of unalloyed copper in
chloride media-a critical review, Corros. Sci. 46 (2004) 109-135.
https://doi.org/10.1016/S0010-938X(02)00257-3

[18] R. Cimpoesu, P. Vizureanu, I. Stirbu, A. Sodor, G. Zegan, M. Prelipceanu, N.
Cimpoesu, N. Ioanid, Corrosion-resistance analysis of HA layer deposited through
electrophoresis on Ti4Al4Zr substrate, Appl. Sci. 11 (2021) 4198.
https://doi.org/10.3390/app11094198

[19] S. Montecinos, P. Klímek, M. Sláma, S. Suarez, S. Simison, Corrosion behavior of a
β CuAlBe shape memory alloy containing stress induced martensite, Appl. Surf.
Sci. 466 (2019) 165-170. https://doi.org/10.1016/j.apsusc.2018.10.047

[20] M.G. Zaharia, S. Stanciu, R. Cimpoesu, I. Ionita, N. Cimpoesu, Preliminary results
on effect of H_2S on P265GH material for gas and petroleum transport, Appl. Surf.
Sci. 438 (2018) 20-32. https://doi.org/10.1016/j.apsusc.2017.10.093

[21] C.P. Liu, S.J. Chang, Y.F. Liu, J. Su, Corrosion-induced degradation and mechanism
study of Cu-Al interface for Cu-wire bonding under HAST conditions, J. Alloys

Compd. 825 (2020) 154046. https://doi.org/10.1016/j.jallcom.2020.154046

[22] H.H. Kuo, W.H. Wang, Y.F. Hsu, Microstructural characterization of precipitates in Cu-10Al-0.8Be SMA, Mater. Sci. Eng. A 430 (2006) 292-300. https://doi.org/10.1016/j.msea.2006.05.061

[23] A.M. Alfantazi, T.M. Ahmed, D. Tromans, Corrosion behaviour of copper alloys in chloride media, Mater. Des. 30 (2009) 2425-2430. https://doi.org/10.1016/j.matdes.2008.10.015

[24] M. Chmielová, J. Seidlerová, Z. Weiss, X-ray diffraction phase analysis of crystalline copper corrosion products, Corros. Sci. 45 (2003) 883-889. https://doi.org/10.1016/S0010-938X(02)00176-2

[25] X. Wang, C. Chen, T. Guo, J. Zou, X. Yang, Microstructure and properties of ternary Cu-Ti-Sn alloy, J. Mater. Eng. Perform. 24 (2015) 2738-2743. https://doi.org/10.1007/s11665-015-1483-4

[26] R.G. Chelariu, M. Benchea, R. Cimpoeşu, O. Rusu, V. Manole, D.P. Burduhos-Negris, N. Cimpoeşu, C. Bejinariu, Structural and mechanical characterization of as-cast CuAlBe alloy, Mater. Today Proc. 72 (2023) 594-599. https://doi.org/10.1016/j.matpr.2022.10.106

[27] R.K. Singh, S. Murigendrappa, S. Kattimani, Investigation on properties of Cu-Al-Be SMA wire doped with zirconium, J. Mater. Eng. Perform. 29 (2020) 11-15. https://doi.org/10.1007/s11665-020-05233-7

[28] A. Hrituc, V. Ermolai, A.M. Mihalache, L. Andrusca, O. Dodun, G. Nagît, M.A. Boca, L. Slatineanu, Compressive behavior of ceramic-polymer composite balls manufactured by 3D printing, Micromachines 15 (2024) 1-10. https://doi.org/10.3390/mi15010150

[29] A. Canbay, Investigation of quenching media effects on thermodynamic and structural features of aged CuAlFeMn SMA, Phys. B Condens. Matter 557 (2019) 117-125. https://doi.org/10.1016/j.physb.2019.01.011

[30] P.M. Wasmer, G. Mussot-Hoinard, S. Berveiller, E. Patoor, A. Eberhardt, Kinetics of precipitation and mechanical behavior of CuAlBe single crystal drawn-wires, ESOMAT Proc. (2009) 06023. https://doi.org/10.1051/esomat/200906023

[31] A.C. Canbay, A. Aydoğdu, Thermal analysis of Cu-14.82Al-0.4Be SMA, J. Therm. Anal. Calorim. 113 (2013) 731-737. https://doi.org/10.1007/s10973-012-2792-6

[32] HG 1058/2006 privind cerinţele minime pentru securitatea şi sănătatea lucrătorilor expuşi la atmosfere explozive.

[33] HSG 103, Manual explosion protection, HSE Books, 2003, ISBN 978-0-7176-2726-4.

[34] A. Hautcoeur, A. Eberhardt, E. Patoor, M. Berveiller, Thermomechanical behaviour of monocrystalline Cu-Al-Be SMAs, J. Phys. IV 5 (1995) 459-464.

[35] R.G. Chelariu, G. Badarau, N. Cimpoesu, C. Bejinariu, A concise analysis of

regulations on intervention and rescue activities at industrial establishments, Bull. Inst. Politeh. Iaşi 69 (2023) 211-220.

[36] G.V. da Mota Candido, D.F. de Oliveira, I.C.A. Brito, R.E. Caluête, B.H. da Silva Andrade, D.G. de Lima Cavalcante, Effect of hot rolling on thermomechanical properties of Cu-Al-Be-Cr alloy, Mater. Res. 23 (2020) 20190542. https://doi.org/10.1590/1980-5373-mr-2019-0542

[37] M. Bansal, N. Sindhu, S. Anand, Structural and model analysis of a composite material differential gearbox assembly, Int. J. Eng. Sci. Res. Technol. (2016) 1-6.

New Shock-Resistant Materials for Work Equipment used in Potentially Explosive Atmospheres
Materials Research Foundations **186** (2026)   https://doi.org/10.21741/9781644903872

CHAPTER 4

Analysis of CuTi Alloys Proposed as an Alternative to Existing Non-Sparking Alloys

Romeo-Gabriel CHELARIU[1], Ramona Cimpoeșu[1]*,Gabriel-Dragoș VASILESCU[2],Costică Bejinariu[1,3]

[1]Faculty of Materials Science and Engineering, "Gheorghe Asachi" Technical University of Iasi, 67 Dimitrie Mangeron Street, 700050 Iasi, Romania

[2]National Institute for Research and Development in Mine Safety and Protection to Explosion— INSEMEX, 332047 Petrosani, Romania

[3]Academy of Romanian Scientists, Ilfov 3, 050044 Bucharest, Romania

ramona.cimpoesu@academic.tuiasi.ro

Abstract

Two experimental Cu-Ti alloys with very good chemical and structural homogeneity were obtained by conventional casting. For the Cu2Ti alloy, a good solubilisation of titanium in the copper matrix and the formation of a solid solution-based eutectic and β-TiCu4 phase were observed. These alloys have demonstrated success in wear tests involving 16,000 stresses in two types of explosive atmospheres, with no occurrence of hot sparks or subsequent explosive explosions. It has been observed that increasing the mass percentage of Ti in copper from 2 to 3% results in an enhancement of both hardness and the modulus of elasticity. This phenomenon is primarily attributed to the formation of titanium-based compounds, particularly TiCu4.

Keywords

Cu2Ti Alloy, Structural Analysis, TiCu4, Microhardness, Coefficient of Friction

4.1 Structural, chemical and phase analysis of experimental CuTi alloys

The maximum solubility of titanium in copper at 885°C is 8% by weight. The intermediate metastable phases with ordered structure may originate before the precipitation of the equilibrium β-TiCu$_4$ phase. Peritectic reactions occurring between the melting point of TiCu phase particles and the eutectic transformation temperature lead to the possibility of the intermetallic Ti$_3$Cu$_4$, Ti$_2$Cu$_3$, TiCu$_2$ and TiCu$_4$ phases formation. Furthermore, two types of TiCu$_4$ equilibrium phases occur in CuTi alloys: one stable and one metastable. The α-phase TiCu$_4$ can be transformed to the β-phase TiCu$_4$ after a prolonged aging treatment, even at low temperatures. Furthermore, the reverse transformation of the β-phase to the α-phase is possible during cooling, provided that low solidification/cooling temperatures are employed [1,2].

As demonstrated in prior investigations [3], CuTi alloys have the potential to undergo precipitation-hardening during the aging heat treatment process, a phenomenon that occurs through the spinodal decomposition mechanism. This mechanism serves to determine the clustering and ordering of the structure. Extensive investigations into the precipitation mechanism in Cu-Ti alloys [4,5] have indicated that the intermetallic β' phase ($TiCu_4$) contributes to matrix hardening during the aging annealing heat treatment. Furthermore, the long-term aging of CuTi alloys has been observed to result in cell precipitation along the matrix boundaries and the subsequent formation (precipitation) of β-$TiCu_3$ equilibrium phases [6,7]. The formation of fine-scale coherent/semi-coherent D1a precipitates ($TiCu_4$) at high supersaturation provides these alloys with strength levels comparable to Cu-Be alloys. Quasi-periodic lattices of D1a particles or modulated structures undergo size augmentation in accordance with a generalized LSW kinetic law, exhibiting an activation energy of approximately 50 kcal mol^{-1}. This finding aligns with the reported values for the activation energy of Ti in Cu.

The initial decomposition stages of supersaturated CuTi alloys involve a complex interplay of ordering and grouping effects in solution. The formation of coherent two-phase mixtures is likely to involve nonclassical nucleation or spinodal decomposition within the framework of a generalized theory of nucleation, accompanied by a precise specification of the spinodal process.

The analyses were performed on the Cu-3(wt%)Ti alloy, hereafter labeled Cu3Ti. A review of the available literature has identified eight intermetallic compounds ($TiCu_4$, $TiCu_3$, $TiCu_2$, Ti_2Cu_3, Ti_3Cu_4, $CuTi$, $CuTi_2$ and $CuTi_3$) in the Cu-Ti system (Taguchi, 1990). It is noteworthy that the majority of prior studies have been conducted on Cu-Ti diffusion alloys. In contrast, Dziadon et al. obtained four distinct intermetallic compounds ($TiCu_4$, Ti_3Cu_4, $CuTi$, and $CuTi_2$) from a Cu-Ti diffusion couple that was maintained below 890 °C for 5-40 minutes. It has been observed that the microhardness of the diffusion zone was 510-550 HV, which is much higher than that of pure copper (72 HV) or pure titanium (220 HV) and attributed this increase to the formation of new intermetallic phases or compounds in addition to the solid solution of copper with the solubilized titanium in its structure (Dziadon, 2004). As illustrated in Fig.s 1(a) and (b), the X-ray diffraction (XRD) spectra were obtained from the Cu3Ti cast and rolled samples. The analysis identified several peaks characteristic of copper and the solid solution that copper forms with titanium (solubilized in the Cu matrix), as well as a peak characteristic of $TiCu_4$ and $TiCu_3$ compounds, the only compounds that can be formed for this chemical composition according to the phase, mass, or forming enthalpy variation diagrams discussed in Chapter 3 of the thesis. A low-intensity peak identified on the surface of the CuTi cast alloy was attributed to titanium dioxide [8].

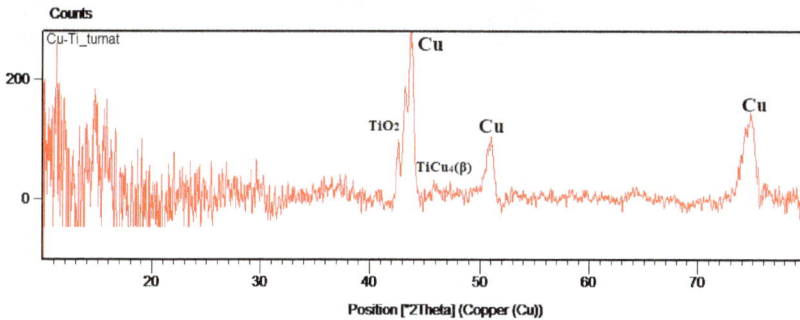

(a)

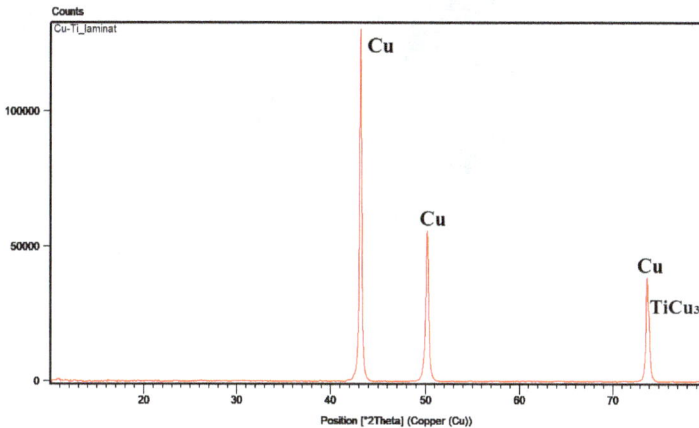

(b)

Figure 1. Characteristic XRD spectra of Cu3Ti alloy a) cast condition and b) hot rolled (900°C) condition.

The diffraction patterns of these zones in the rolled sample stage indicate only the occurrence of periodic stresses in the crystal lattice, which were induced by coherent particles of intermetallic phases and re-ordering of the structure in the more titanium-rich zones. This re-ordering is only possible by achieving the relevant chemical composition. The microstructure morphology of the main cast alloys is homogeneous dendritic, as illustrated in Fig. 2(a). The employment of a steel casting mold allowed improved alloy homogenization, resulting in a random distribution of titanium in the copper matrix. The dominant morphology in the microstructure is dendritic. The presence of a substantial equilibrium β phase within the structure is attributed to the comparatively moderate solidification rate.

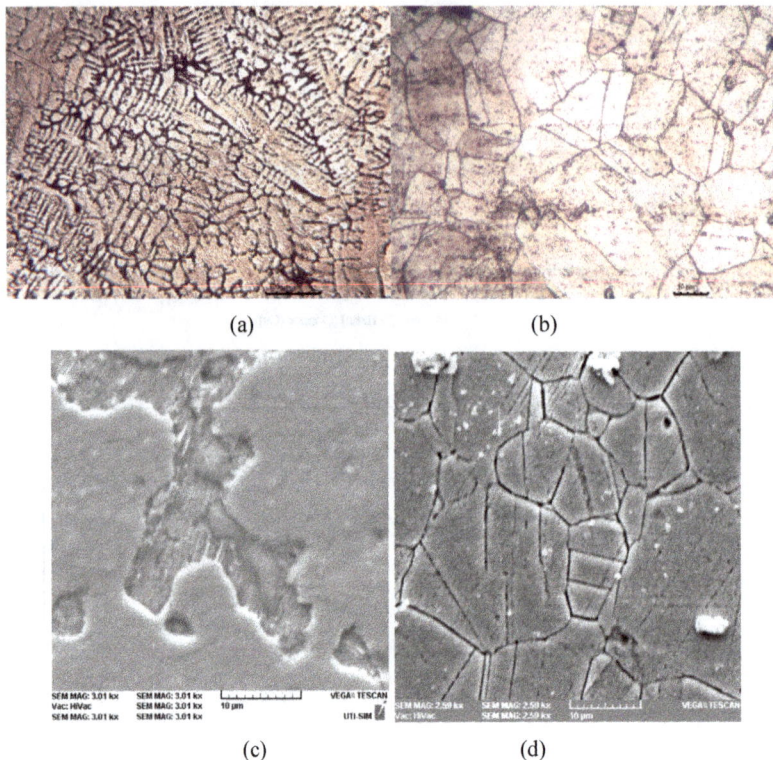

(a)   (b)

(c)   (d)

Figure 2. Optical microscopy analysis of Cu3Ti alloy (a) cast, (b) rolled and electron microscopy (c) cast, (d) rolled.

The equilibrium phase in the Cu-Ti system is typically comprised of classical Widmanstatten or cell precipitation. The cellular or "discontinuous" precipitation reaction is a central component in the overaging of these high-strength alloys. The growth kinetics of the cellular microconstituent are governed by an activation energy that is less than half the activation energy for lattice diffusion of Ti into Cu. A defining feature of these alloys is the intermetallic lamellar distribution, as well as the gradient along the Cu/Ti chemical potentials, which dictates that the reactions are governed purely by solid-phase diffusion [9]. Furthermore, due to the low diffusion rate, the thicknesses of each lamella are limited to a few microns, as illustrated in Fig. 2(c), even if the diffusion time is sufficiently long [9].

It has been reported that the equilibrium phase in hardened Cu-Ti alloys has the composition $TiCu_4$ and may exhibit polymorphic transformation. The high-temperature phase has been identified as the Au_4Zr-type orthorhombic phase, classified by the space group Pnma. In contrast, the low-temperature polymorph has been determined to be the tetragonal D1a phase. Microscopically, Fig. 2(c), as a result of discontinuous transformation, equilibrium phase lamellae appear in the Cu3Ti alloy arranged alternately with the solid solution lamellae. The coherent precipitates, which are formed as a result of spinodal transformation, are periodically arranged. Spinodal decomposition is a process by which a single thermodynamic phase spontaneously separates into two phases without the need for a nucleation event. This decomposition occurs when there is no thermodynamic barrier to phase separation. Consequently, phase separation by decomposition does not require nucleation events resulting from thermodynamic fluctuations, which normally trigger phase separation. It is important to note that equilibrium phases are classified into two distinct categories:

- the non-ordered, high-temperature phase, β'

- the low-temperature ordered phase, β.

The β' phase crystallizes in the hexagonal lattice, while the β phase in the rhombic lattice [10].

It is imperative to employ a plastic deformation temperature that is lower than the solid solution formation and recrystallization temperature in order to facilitate the precipitation of a fine and uniformly distributed secondary phase within the main phase. Conversely, if the annealing temperature exceeds the solid solution formation temperature, no secondary phase precipitates in the main phase. The fine and uniform secondary phase formed by intermediate annealing in the main phase contributes to preventing coarsening of the crystalline grains in the main phase during the solution heat treatment and hence to the development of a heat treated structure with an average crystalline grain size not exceeding 25 microns, Fig. 2 (d). The distribution of the constituent elements indicates that the phase containing the greatest proportion of titanium is also the least prevalent, manifesting as a eutectic in dendritic form (Fig. 3 (a)-(d).

The hot deformation of the alloys was performed on the as-cast samples. A total of four to five deformation passes were conducted at low degrees (less than 10%) following the ingots' heating at 900-950°C for 10 minutes. As the deformation degree increases, the samples transition from uniform to non-uniform deformation. In the uniform deformation stage, the number of twinned boundaries clearly increases (see Fig. 2(b)). In the non-uniform deformation stage, additional slip bands perpendicular to the rolling direction emerge (Fig. 2(d)).

(a)   (b)

(c)   (d)

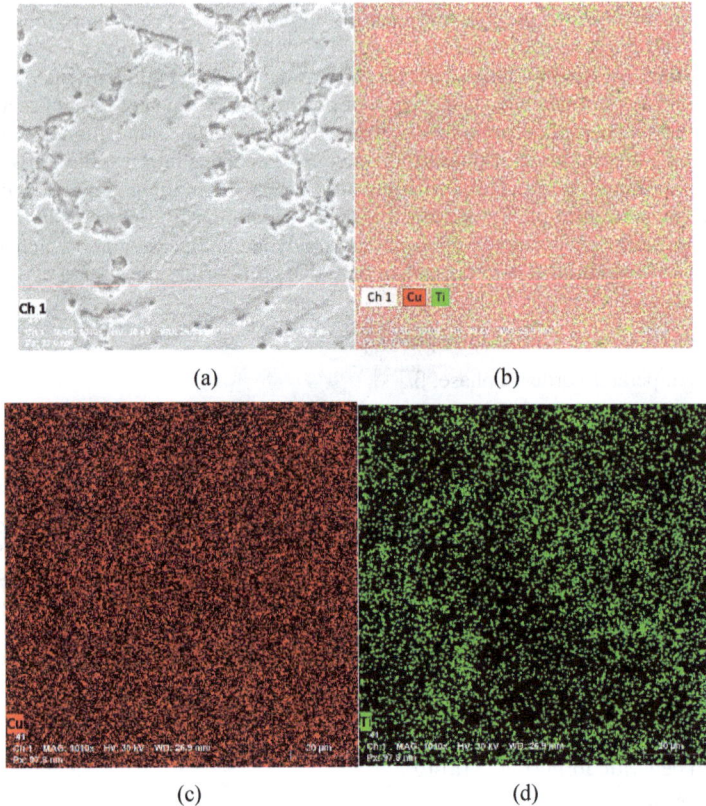

Figure 3. Main Cu and Ti element distribution in the cast alloy (a) area selected for analysis, (b) overlapped distribution of Cu and Ti elements, (c) Cu distribution and (d) Ti distribution.

The process of hot rolling, in conjunction with an alloy ageing treatment, has been observed to induce the dissolution of precipitated second-phase TiCu4 particles. This phenomenon has been shown to result in a modest hardness reduction, contrary to the usual rolling hardening behaviour (based on two phenomena: strain hardening and grain boundary growth) [11].

The experimental alloys, Cu2Ti and Cu3Ti, were obtained from high-purity materials, and no impurities were detected by X-ray diffraction or energy dispersive X-ray spectroscopy analysis (Fig. 4 (b)). In addition to tests determining the chemical composition of the alloys on 4 mm2 areas (3 determinations for each sample), analyses were also performed at various points on the alloy surface, Fig. 4 (a) for the cast alloy and Fig. 6 (a) for the rolled alloy. The qualitative analysis showed that only the two main constituent elements of the alloy, copper and titanium, were identified, Fig. 4 (b).

The consolidation of deformed alloys by increasing the number of grain boundaries and the multiplication of coherent titanium-rich precipitates is accelerated by increased diffusion. A similar phenomenon was previously documented when the degree of cold plastic deformation was increased from 0% to 80%, resulting in an enhancement of surface hardness [12].

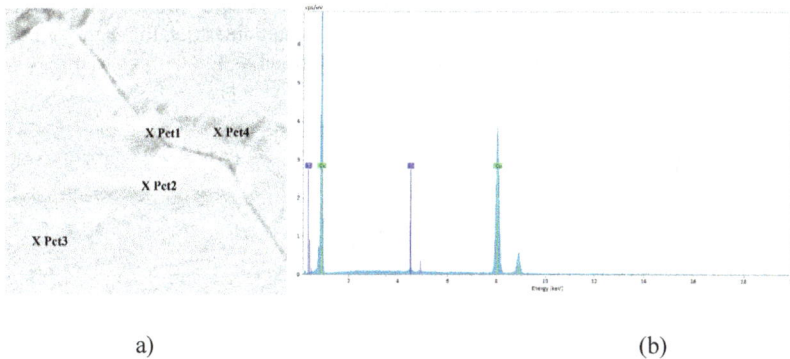

a)   (b)

Figure 4. Selection of points for the chemical analysis of Cu3Ti alloy in the as-cast state in (a) and the energy spectrum of the chemical elements identified in the alloy in (b).

The chemical compositions as determined by EDS analysis (Table 1) confirm the proposed chemical compositions of the Cu2Ti and Cu3Ti alloys. Following hot rolling, a slight reduction in the mass percentage of Ti was observed, from 2.9 to 2.7. The copper matrix solubilises various amounts of titanium, with mass percentages determined between 0.9 and 2.6 Ti in the case of the cast alloy (Cu2Ti) and 2.3- 3.7 in the case of the rolled Cu3Ti alloy.

Table 1. Chemical composition of the selected elements in Fig. 4 (a) for the cast alloy and in Fig. 6 (a) for the rolled alloy.

Alloy condition	Area Element		General	Point 1	Point 2	Point 3	Point 4	EDS Error %
Cast	Cu	wt	97.9	93.3	98.7	99.1	97.4	1.5
		at	97.3	91.3	98.2	98.8	96.6	
	Ti	wt	2.1	6.7	1.3	0.9	2.6	0.2
		at	2.7	8.7	1.7	1.2	3.4	
Rolled	Cu	wt	97.4	97.5	96.3	97.7	85.8	2
		at	96.5	96.7	95.2	96.9	82.1	
	Ti	wt	2.65	2.5	3.7	2.3	14.2	0.2
		at	3.5	3.3	4.8	3.1	18.0	

St.Dev.: Cu±1.2; Ti:±0.1

As illustrated in Fig. 6, the presence of the TiCu4 compound (with 18at%) was identified for the Cu3Ti alloy. This finding aligns with the results reported in the existing academic literature as well as the results previously outlined in the XRD analysis section.

Furthermore, the presence of compounds exhibiting a higher percentage of titanium was observed and confirmed through variation tests of the chemical composition of the two primary elements, copper and titanium, on a 40 µm line (Fig. 5 (a) and (b)). The distribution of titanium was found to be homogeneous within the matrix. The term 'main or matrix phase' as used here means the α-phase in a binary phase diagram for a copper-titanium alloy, and 'secondary phase' means the precipitate of an intermetallic compound, e.g. $TiCu_4$. The solid solution formation temperature is defined as the temperature that marks the boundary between the „α+ Ti Cu₄" phase and the α phase [13].

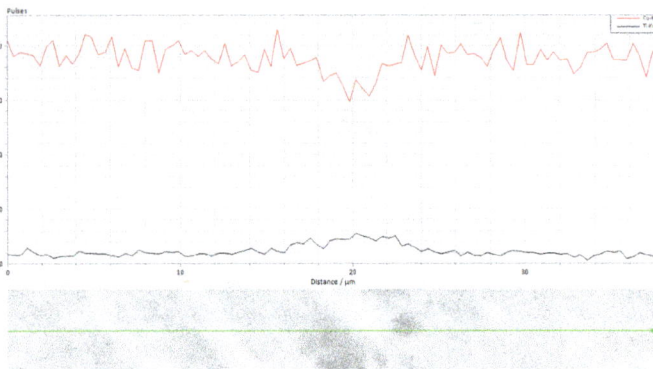

(a)

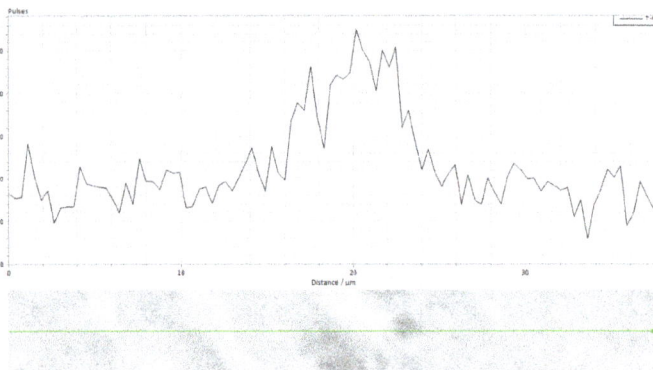

(b)

Figure 5. Concentration variation of Cu and Ti elements over a distance of 40 µm in (a) and only Ti in (b) for the cast alloy.

The elemental distributions exhibited on the surface, as depicted in Fig. 6, reveal the formation of compounds that contain elevated concentrations of titanium, as illustrated in Fig. 6(d). This observation is in contrast to the solubilised Cu+Ti matrix.

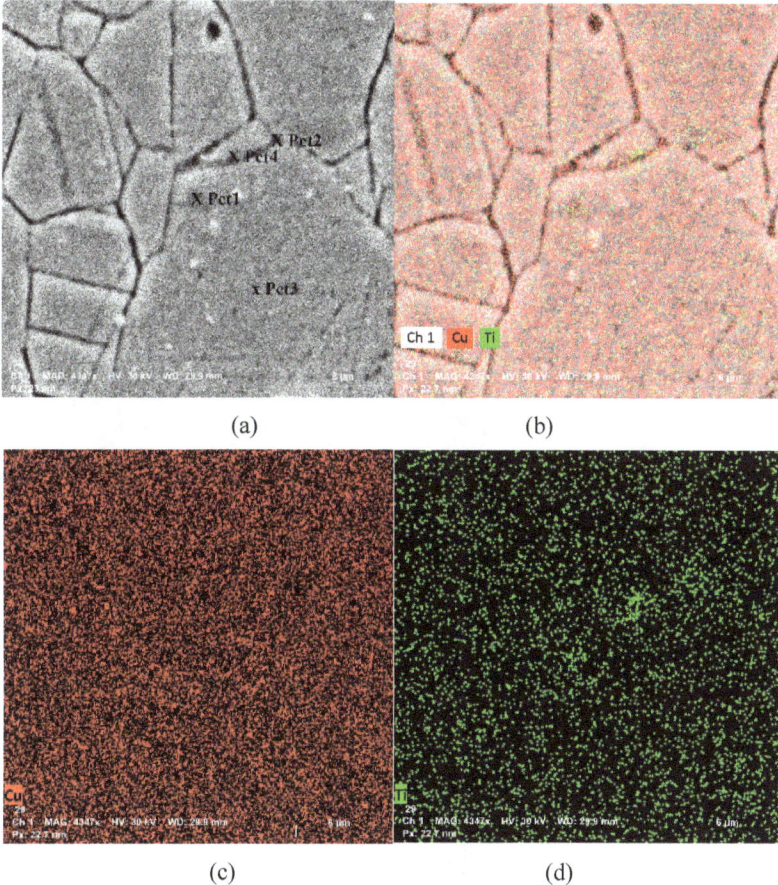

(a)   (b)

(c)   (d)

Figure 6. Main Cu and Ti component elements distribution in the rolled alloy (a) the surface selected for the distribution analysis of the main chemical component elements and the 4 zones selected for the point analysis of the alloy, (b) Cu and Ti elements distribution on the selected surface, (c) Cu distribution and (d) titanium distribution.

In order to corroborate and emphasise the formation of intermetallic compounds that contain a greater proportion of titanium, analysis was conducted to ascertain the distribution of component elements on a line (Fig. 7 (a) and (b)).

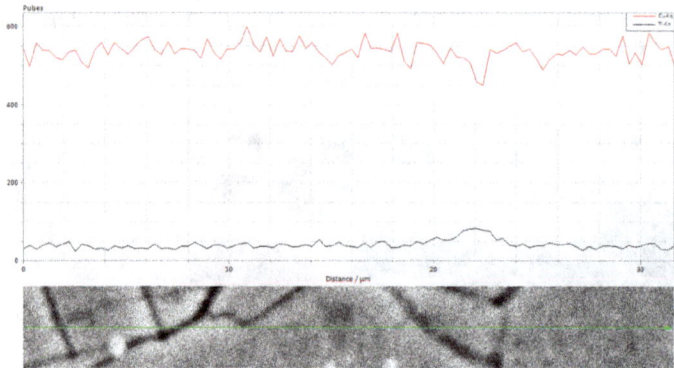

(a)

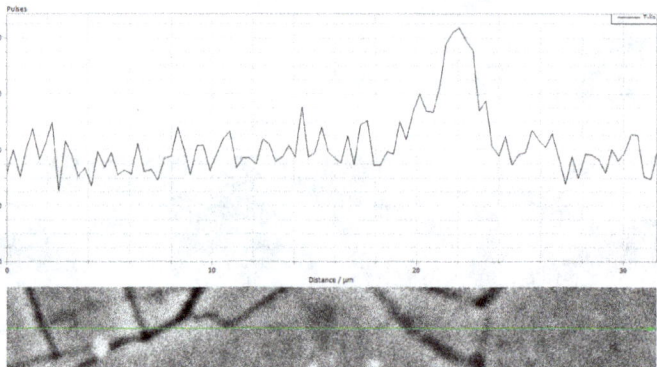

(b)

Figure 7. Concentration of Cu and Ti elements over a distance of 30 μm in (a) and only Ti in (b) for the rolled alloy.

Cu-Ti alloys that have been subjected to plastic deformation are also exposed to an intermediate annealing heat treatment at a temperature that is lower than the solid solution forming temperature. This is to ensure the successful distribution and precipitation of a secondary phase within a main phase or matrix.

4.2 Electrochemical corrosion resistance of CuTi alloys

Copper titanium (Cu-Ti) alloys have been the focus of extensive research in the Cu-rich area, which is regarded as a promising substitute for the widespread Cu-Be alloy for electronic components such as connectors and relay control elements [8] or non-sparking elements. A substantial number of investigations have been reported on the effect of Ti content and heat treatment conditions on electrical conductivity, thermal conductivity and mechanical properties [14]. In contrast, comparatively fewer investigations have focused on CuTi intermetallic compounds, although pioneering studies have demonstrated the potential application of these alloys in high conductivity cables, grinding wheels, and marine components [15]. Furthermore, there have been only a limited number of studies conducted on the corrosion resistance of these alloys [8].

The experimental tests were conducted in a 3.5% NaCl saline solution, which was prepared in the laboratory to analyse the behaviour in possible environments with salty atmospheres or in various electrolyte solutions with high salinity, such as the marine environment. The potentiodynamic polarization curves of Cu-3%wt Ti in saline electrolyte solution are presented in Fig. 8 and the corrosion current densities and corrosion potentials obtained by Tafel extrapolation are given in Table 2. A thorough analysis of the curves reveals that there is not a significant disparity between the behaviour of the two distinct alloy states in terms of corrosion resistance, as the curves exhibit substantial overlap.

Table 2. Characteristic electrochemical parameters of the linear potentiometric test (with Tafel interpretation).

Sample	$E(I=0)$ (mV)	i_{corr} (μA $/cm^2$)	Rp kohm.cm	V_{corr} ($\mu m/y$)	$-\beta_c$ (mV/dec)	β_a (mV/dec)
Cu3Ti Cast	-287.5	14.5790	1.60	168.5	239.1	128.0
Cu3Ti Rolled	-319.6	15.0904	2.54	174.4	265.5	157.2

As demonstrated in Table 2, the value of the cathode reactions, $-\beta_c$, is higher than that of the anode (β_a) due to a higher intensity of the reduction reactions. Given the noble nature of Cu, it is the most common cathode in contact with other metallic materials and therefore the cathodic properties of Cu are very important. The oxygen reduction reaction, facilitated by a four-electron mechanism, is the most significant cathodic reaction in the context of the corrosion of pure Cu or alloys comprising more than 95% Cu:

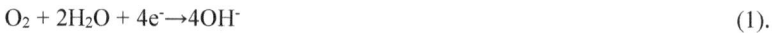

$$O_2 + 2H_2O + 4e^- \rightarrow 4OH^- \qquad (1).$$

The corrosion current for both states of the alloy is higher than that reported by Ochoa et al. for pure Cu in a 0.5 M NaCl solution [16,17].

In an aqueous solution, Cu(II) and Cu(I) are the stable states, with the latter being only slightly soluble in water, resulting in the predominant formation of a Cu_2O film during the corrosion process. Conversely, Cu^{2+} is the predominant soluble species. The solubility of Cu(I) is enhanced in the presence of Cl- or other complexing agents, and it becomes the dominant species in solution, e.g. as $CuCl_2^-$.

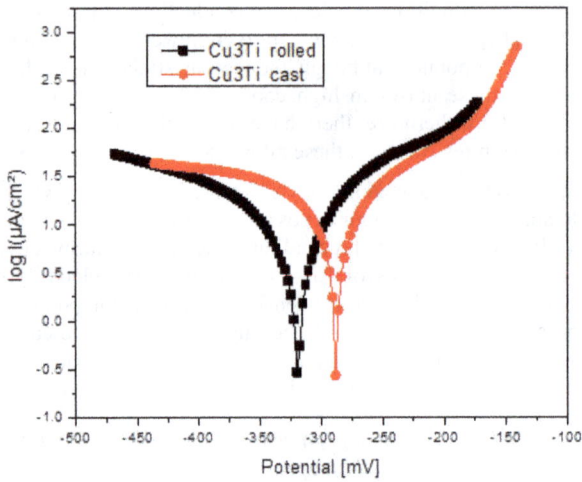

(a)

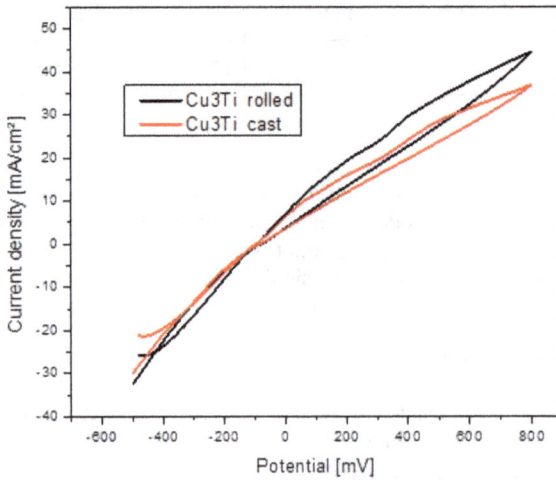

(b)

Figure 8. Electro-chemical corrosion behavior a) Tafel diagram and b) cyclic potentiometry.

As demonstrated in Table 2, the corrosion rate (i_{corr}) of the cast and rolled samples was found to be comparable, with a minor increase observed in the rolled sample, likely attributable to the stresses induced during the rolling process. The E_{corr} corrosion potential values were found to be highly similar, indicating a comparable degree of corrosion resistance. Furthermore, Fig. 8 (b) demonstrates the general corrosion characteristics of CuTi alloys in salt solutions, exhibiting a moderately elevated current density in the rolled alloy relative to the cast one. This generalised corrosion behaviour was substantiated through subsequent optical and electron microscopy analyses.

As illustrated in Fig. 9, optical microscopy was employed to observe the surfaces of the experimental samples during the process of electrochemical corrosion resistance testing. It was found that the area exposed to the electrolyte solution resulted in the formation of a layer of dark green corrosion compounds, which spread across the entire surface in contact with the electrolyte solution.

The presence of copper chloride (CuCl) in combination with copper oxide (Cu_2O) and hydroxide ($Cu(OH)_2$) is indicated by the presence of small, low percentage, blue-coloured areas on the micrographs. Studies have shown that in neutral chloride solutions, the dominant corrosion product on the copper surface is CuCl, which eventually converts to Cu_2O, which oxidises to $Cu(OH)_2$, and $Cu_2(OH)_3Cl \times Cu(OH)_2$ (Kear, 2004).

(a)   (b)

Figure 9. Optical microscopy of the surfaces after the electrochemical corrosion resistance test a) as cast and b) as rolled.

As demonstrated in Figs. 9 and 10, the layer formed by surface corrosion was not removed by ultrasonic cleaning and exhibited high adhesion to the substrate. Analysis of the SEM images in Fig. 10 indicates that the layer formed on the surface is not a compact layer, but consists of several compounds grown from the interaction of the CuTi alloy with the electrolyte solution (saline water).

(a)   (b)

(c)   (d)

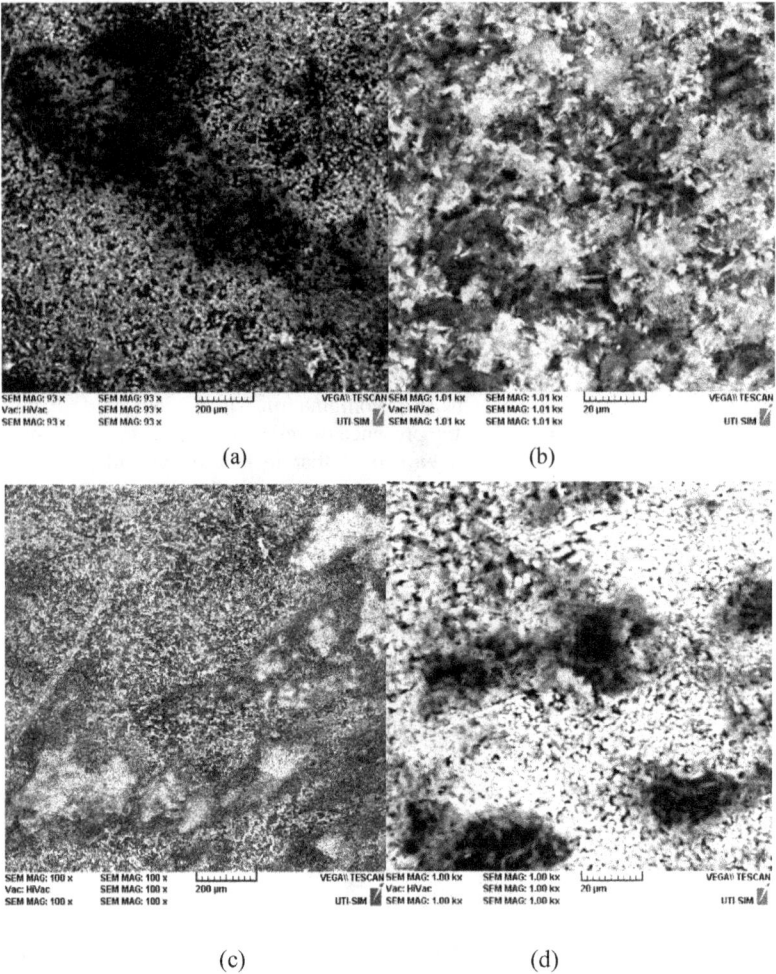

Figure 10. SEM images of the experimental alloys after the electrochemical corrosion resistance test of the cast alloy: a) 100x and b) 1000x and of the rolled alloy: c) 100x and d) 1000x.

The surface was found to be covered with corrosion products, the composition of which was primarily chloride based, as observed in Fig.s 11, 13 and 14. This is attributed to the use of a chloride-rich electrolyte, containing high concentrations of NaCl-based components. The corrosion exhibited uniform distribution, with both surfaces being thoroughly coated by corrosion products. Localised corrosion was not observed on the surface (Fig.s 9 and 10).

EDS analysis was performed to identify the elements that comprise the layer of corrosion compounds, which was found to consist of Cu, Ti, O, Cl and Na. This analysis was conducted both qualitatively, as illustrated in Fig. 11, and quantitatively, as outlined in Table 3.

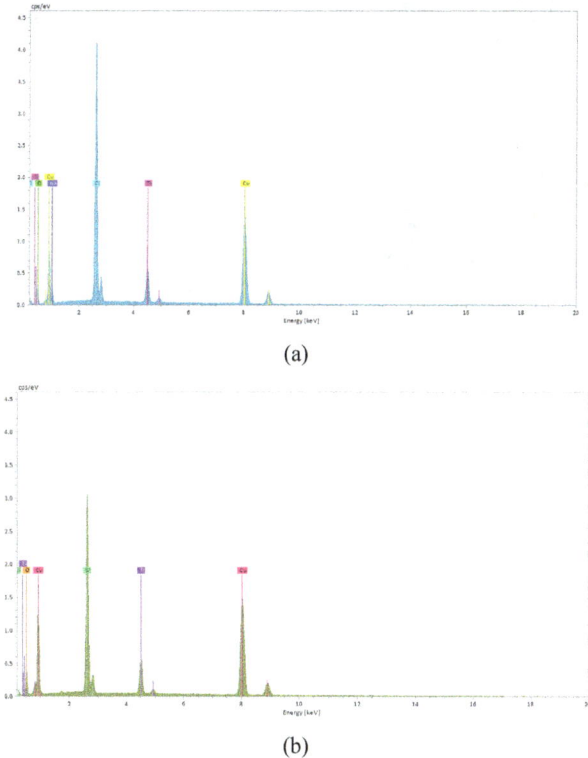

(a)

(b)

Figure 11. Energy spectra recorded on the surface of the corroded alloy (a) point 1 and (b) points 2 and 3 in Fig. 12a).

The corrosion morphology can be explained by considering the presence of $TiCu_4$ precipitates. These precipitates were more noble than the Cu matrix and, when corrosion occurred, the matrix corroded first. In order to explain the corrosion mechanism, it is also necessary to consider the EDS mapping analysis shown in Fig. 13 and 14.

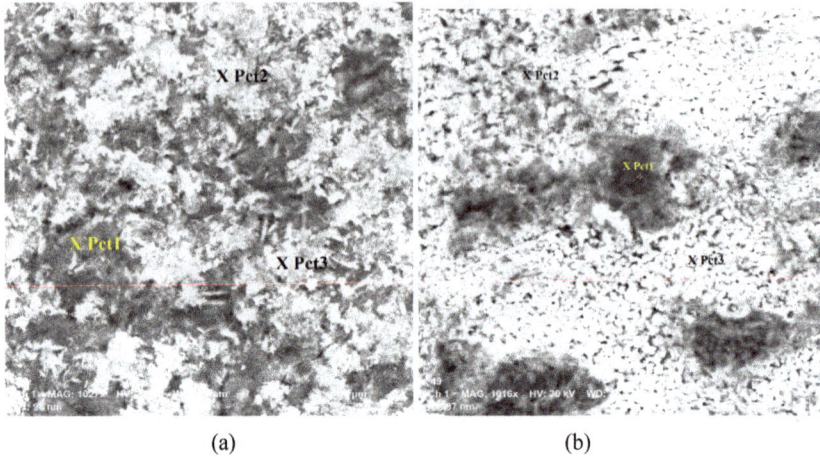

(a)   (b)

Figure 12. Sample surface after electrochemical corrosion resistance test and selection of chemical analysis points (a) cast and (b) rolled.

As demonstrated in Fig. 12(a), the cast sample exhibited regions with elevated precipitate concentrations, coexisting with other regions deficient in precipitates. In contrast, the rolled sample (Fig..12(b)) demonstrated a uniform distribution of precipitates within the Cu matrix. The existence of regions characterised by substantial concentrations of noble precipitates, exhibiting distinct corrosion potentials relative to the Cu matrix, gives rise to the formation of a galvanic couple. It is noteworthy that areas exhibiting high concentrations of Ti precipitates are observed to possess a greater degree of nobility in comparison to those areas devoid of precipitates. The cast sample demonstrated superior corrosion resistance, attributable to the immediate corrosion of the Cu matrix, precipitated by the galvanic coupling that formed between the distinct zones. The oxide layer exhibited a reduced time to form on the surface. Chlorine and oxygen-based compounds were identified on both the cast and rolled samples, while sodium-based compounds were detected on the rolled sample (Table 3 and Fig.s 13 and 14).

Table 3. Chemical compositions obtained from the corroded surfaces of the experimental alloys on large areas and selected points in Fig.s 12 (a) and (b).

Element / area	Cu %		Cl %		O %		Ti %		Na %	
	wt	at	wt	at	wt	at	wt	at	wt	at
Cast point 1	40	21.3	29.3	28.2	16.1	34.3	7.7	5.5	7.2	10.8
Cast point 2	48.1	26.5	24.6	24.1	20.1	44	7.43	5.4	-	-
Cast point 3	47.8	26.2	23.2	22.7	20.7	45.1	8.35	6.1	-	-
General cast	42.6	22.4	25.8	24.4	22.3	46.6	9.4	6.5	-	-
Rolled point 1	42.3	23.1	26.4	25.8	17.8	38.6	10.1	7.3	3.5	5.2
Rolled point 2	50.2	28.4	26.1	26.4	17.5	39.2	4.7	3.5	1.6	2.6
Rolled point 3	56.3	36.6	32.4	37.8	9.4	21.6	1.5	1.3	1.5	2.7
General rolled	49	27.5	26.6	26.7	17.3	38.6	4.8	3.6	2.3	3.5
Detector EDS Error %	1.3		0.9		2.1		0.3		0.3	

With regard to the degree of oxidation, as illustrated in Table 3 (general chemical composition), the cast sample demonstrates a higher degree of oxidation. Copper(II) chloride, also referred to as cupric chloride, is an inorganic compound with the chemical formula $CuCl_2$. The yellowish-brown monoclinic anhydrous form (Fig. 9) gradually absorbs moisture, resulting in the formation of the blue-green orthorhombic dihydrate $CuCl_2 \cdot 2H_2O$, which contains two hydration water molecules. As demonstrated in Table 5.3, the results of point analyses reveal the presence of areas, such as point 1 on the rolled sample (Fig. 12 (b)), where titanium occurs in a higher percentage. This is likely attributable to the presence of TiCu-based compounds that have undergone slight oxidation but have not formed compounds with chlorine or sodium, nor been coated by these compounds on the surface.

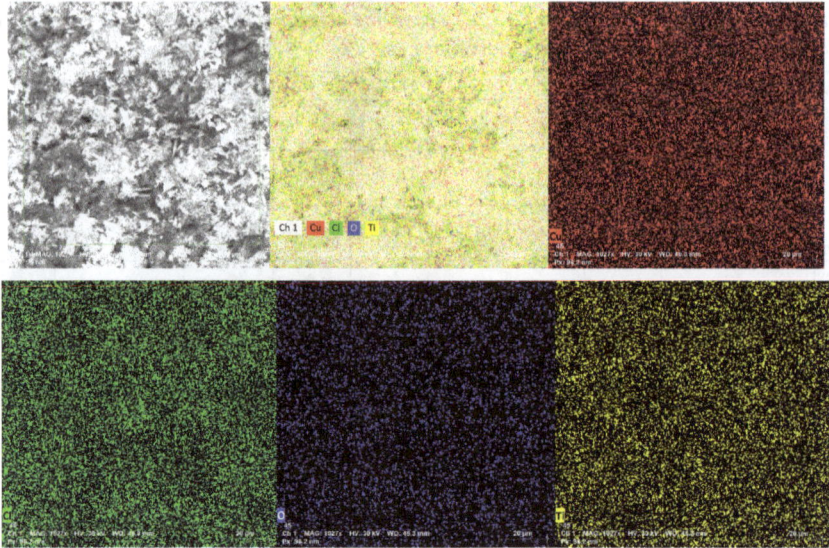

Figure 13. Main element distributions identified on the surface of the cast alloy after corrosion.

It has been demonstrated that copper can be dissolved from a solution of $NaCl/O_2/H_2O$. Oxygen is known to oxidise copper over a long period of time, and NaCl has been shown to lower the electrochemical potential for this reaction, probably forming some complex chloride compounds such as $[CuCl_2]^-$, $[CuCl_4]^{2-}$ etc. Theoretically, in the absence of oxygen, no reaction would be expected to occur. The stability of complex compounds is directly proportional to the reduction in potential. Consequently, when the system finally 'chooses' the most stable complex possible (in this case, - $CuCl_4(^{2-})$), the reaction potential of $Cu(0)$ $^{-2e}$ + $4Cl(^-)$ —> $CuCl_4(^{2-})$ will be significantly lower than that of oxygen: $O_2+4e +2H_2O$ —>$4OH(^-)$.

As demonstrated in Fig. 14, the regions exhibiting elevated concentrations of titanium (Ti) exhibit a decline in chlorine (Cl) and an increase in oxygen (O) levels. This phenomenon can be attributed to the formation of titanium oxide (TiO2) on the surface of these materials.The presence of a layer comprising compounds derived from copper chlorides, titanium oxides, and salts has been shown to impede the corrosion process of Cu-Ti based alloys in saline solutions, at least to a certain extent and for a designated duration. The higher presence of Na-based compounds in the rolled sample compared to the cast sample, where they were only locally identified, may be related to the size of the compounds forming in the CuTi alloy.

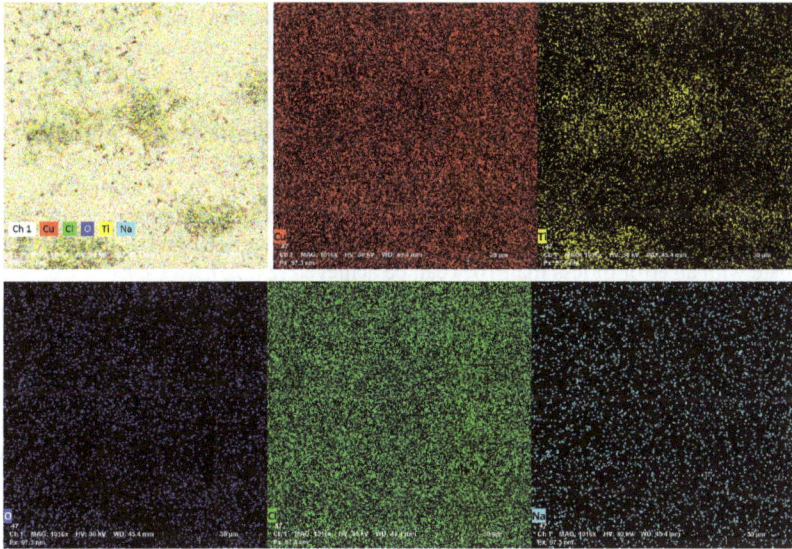

Figure 14. Main element distributions identified on the rolled alloy surface after corrosion.

4.3   Mechanical properties of Cu2(3)Ti alloy

The process of alloying copper with other elements, in this case titanium, is also employed to modify the mechanical properties of pure Cu, which are relatively low. For alloys proposed for industrial applications other than electrical or electronic, the alloying of copper is imperative. In this subchapter, the mechanical characteristics of Cu2Ti and Cu3Ti alloys have been examined, as well as the surface condition of the alloys after hot spark tests in explosive atmospheres.

4.3.1   *Microhardness and coefficient of friction analyses of Cu2Ti and Cu3Ti alloys*

The indentation test was originally developed for the purpose of quality control of materials, utilising the concept of hardness proposed by Brinell as the ratio of applied force to effective contact area. Advances in technology have enabled the enhancement of test equipment, enabling the continuous and simultaneous recording of indenter force and displacement. The accuracy of displacement measurement has been demonstrated to reach a sub-nanometre scale. The ability to conduct indentation directly on the material surface, obviating the need for sample preparation, has firmly established the indentation test as a leading mechanical testing method for a wide range of applications, including microelectronics, thin films, and coatings. Despite the apparent simplicity of the test, the non-linearity of the contact condition will induce complex mechanical deformation fields and stresses even in the case of linear material behaviour. The calculations for the determination of the mechanical characteristics are based on the rules of plastic flow as proposed in the pioneering work of Prandtl [18,19], and only the unloading part of the load-displacement curve, such as the Oliver-Pharr [20,21] and Doerner-Nix [22,23] methods have been utilised for the estimation of the elastic characteristics. These functions are integrated in the indentation equipment employed for these determinations. The aforementioned two methods are based on the assumption that the onset of discharge is elastic. The aforementioned methods propose different ways of evaluating the needle (contact area observed at maximum load), leading to an efficient and practical way to evaluate the Young's modulus characteristic of solid materials, and in most cases, an estimate with an error of less than 10% can be achieved [24, 25].

In the fields of materials science and solid mechanics, the Poisson ratio (v) is a measure of the Poisson effect, defined as the deformation (expansion or contraction) of a material in directions perpendicular to the specific loading direction. The value of Poisson's ratio is negative of the ratio of transverse strain to axial strain. For small values of these changes, the Poisson ratio is defined as the amount of transverse elongation divided by the amount of axial compression. It is observed that the majority of materials exhibit Poisson's ratio values ranging from 0.0 to 0.5. In the case of soft materials, such as rubber, where the bulk modulus is considerably larger than the shear modulus [26], the Poisson's ratio approaches 0.5. For open-cell polymer foams, the Poisson's ratio approaches zero due to the tendency of cells to collapse under compression. Many typical solids exhibit Poisson's ratios within the range of 0.2 to 0.3. Copper has a Poisson's ratio of 0.33, while titanium has a range of 0.265 to 0.34, also with a value of 0.33 (Young's modulus of indentation calculation).

As illustrated in Fig. 15 (a), a schematic representation of the total mechanical work during the indentation procedure is presented. The segment of the area situated beneath the creep zone, extending from C_{max_0} to C_{max}, quantifies the proportion of energy dissipated during the constant application of the maximum load, Fz_max. The duration of force application is contingent upon the hardness testing procedures, as delineated in Chapter 3.

The load-discharge curves for the investigated alloys exhibited a comparable trend; however, discrepancies in the values obtained were observed, attributable to the response of the alloys. As depicted in Fig. 15 (b), the representative surface areas of the plastic mechanical work exhibited higher values in all cases when compared to those of the elastic mechanical work. With regard to creep rates, it is observed that the values decrease in sequence from Cu2Ti cast to Cu3Ti cast and Cu2Ti rolled. The indicative values extracted from the load-discharge plots are 126 nm at 115 nm and 108 nm, respectively.

Table 4 provide the main values from the various regions of the samples, along with the mean values. These demonstrate that the hardness of the alloys is closely related, with minimal disparities observed between regions or even between distinct alloys. The highest value was recorded for the cast Cu3Ti alloy, while the lowest was observed for the rolled alloy. Larger deviations were observed in the values of Young's modulus of indentation (a measure of the stiffness of an elastic and isotropic material). Firstly, there was an almost twofold increase in the indentation mode by increasing the mass percentage of Ti from 2 to 3 wt%. It was observed that with the incorporation of titanium above 2% (as confirmed by structure and phase tests), a higher number of intermetallic and intercrystalline compounds, such as TiCu4, were formed. This resulted in an increase in the microhardness of the alloy from 1.6 to 1.7 GPa, accompanied by an increase in the indentation modulus from 52 to 96 GPa.

The theory of hardening of alloys by intergrain boundary density [27] asserts that the boundary strength is greater than the intracrystalline strength, thereby impeding dislocation movement during the deformation process. The boundaries between grains represent the aggregation of various crystal defects, which significantly influence dislocation displacement. Consequently, for cast alloys, the collective impact of multiple factors (grain refinement, eutectic formation, and intermetallic compound formation) is predicted to result in an enhancement of the alloy's mechanical strength with an increase in Ti content.

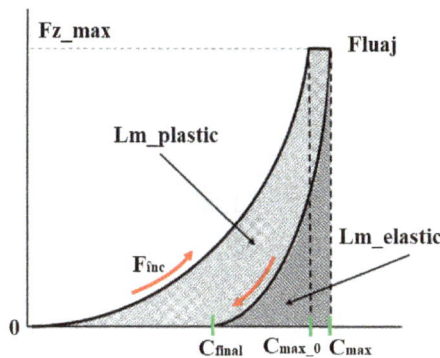

(a)

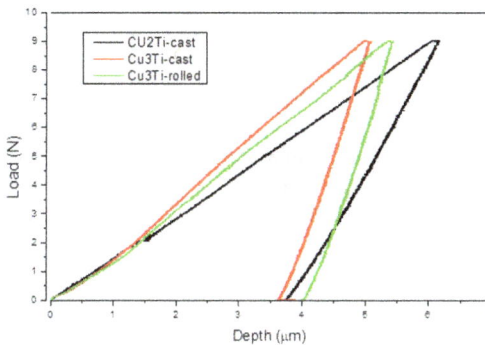

(b)

Figure 15. Schematic representation of the plastic, elastic and creep components for the indentation curves (loading/unloading) as a function of applied force and displacement in (a) and (b) experimental micro-indentation curves of Cu2Ti cast, Cu3Ti cast and Cu3Ti rolled alloys.

The prevailing opinion is that a linear relationship exists between hardness and strength. Consequently, an increase in the titanium content results in a corresponding sharp increase in the hardness of the alloys. However, it is important to note that the enhancement of strength and hardness will inevitably have a detrimental effect on plasticity.

Table 4.4. The main mechanical characteristics extracted from the indentation test (5 indentations were made in different areas on the surface and averaged in the sixth line and the standard deviation is given in brackets).

Alloy /area/ property		Young's indentation modulus (GPa)	Hardness (GPa)	Contact depth	Contact stiffness (N/μm)	Contact area (μm²)
Cu2Ti_cast	1	65.9	1.56	4.7	5.9	5773
	2	40.11	1.7	4.3	3.6	5312
	3	45.14	1.8	4.1	3.9	5072
	4	56.1	1.6	4.44	5	5519
	5	57.3	1.6	4.5	5.1	5547
	M(StDev)	52.91(±9.2)	1.7(±0.1)	4.4(±0.2)	4.7(±0.8)	5445(±237)
Cu3Ti_cast	1	107.5	1.68	4.3	9.1	5371
	2	80.5	1.77	4.1	6.8	5100
	3	73.6	1.93	3.8	5.9	4665
	4	107.7	1.6	4.5	9.3	5633
	5	111.75	1.56	4.7	9.7	5787
	M(StDev)	96.2(±15.9)	1.7(±0.1)	4.3(±0.3)	8.2(±1.5)	5311(±399)
Cu2Ti_rolled	1	116.6	1.59	4.58	10	5679
	2	74.3	1.76	4.1	6.3	5122
	3	115.3	1.52	4.8	10.1	5926
	4	74.2	1.52	4.7	6.7	5909
	5	59.8	1.7	4.3	5.2	5294
	M(StDev)	88(±23.4)	1.6(±0.1)	4.5(±0.3)	7.7 (±2)	5586(±325)

As the degree of rolling deformation increases, the hardness, modulus of elasticity and contact stiffness increase concomitantly with slight variations in contact depth and contact area. This phenomenon is primarily attributable to the generation of a substantial number of deformation structures during the rolling process. Specifically, the grains undergo displacement along the deformation direction, with some grains being crushed, thereby reducing their size to micro-crystalline levels. During the plastic deformation of the grains, the dislocation sources within the grains begin to slip and multiply along a specific crystal plane, thereby generating a substantial number of dislocation centres. The external force-induced accumulation of dislocation zones at the grain boundary has been shown to generate a stress field and activate dislocation sources in adjacent grains, thereby leading to grain refinement. In the rolling deformation process, dislocation motion-induced deformation occurs concomitantly with increasing process strain, and it has been demonstrated that the increase in strain resistance is proportional to the dislocation density.

In the case of the Cu2Ti alloy, in addition to the strain-hardening effect, the precipitated phase and the formed eutectic will split into several parts, which leads to an increase in the stiffness of the alloy. Moreover, certain defects, such as porosity, are known to dissipate following the process of rolling deformation. This process has been shown to enhance the morphology and distribution of inclusions or brittle phases, resulting in a more uniform grain distribution. This, in turn, has been demonstrated to improve the mechanical properties of the material. Scratch tests were conducted to ascertain the coefficient of friction (COF) of the materials under investigation. The results of these tests are illustrated in Fig. 16.

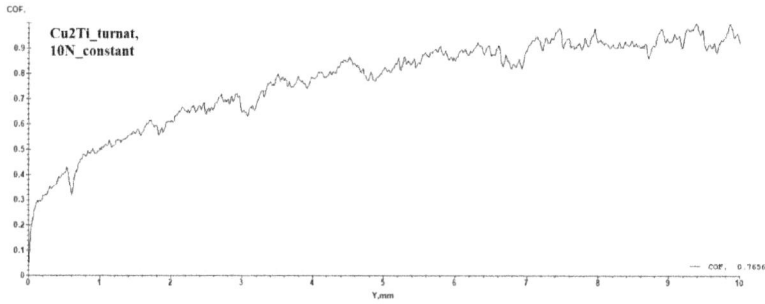

(a)

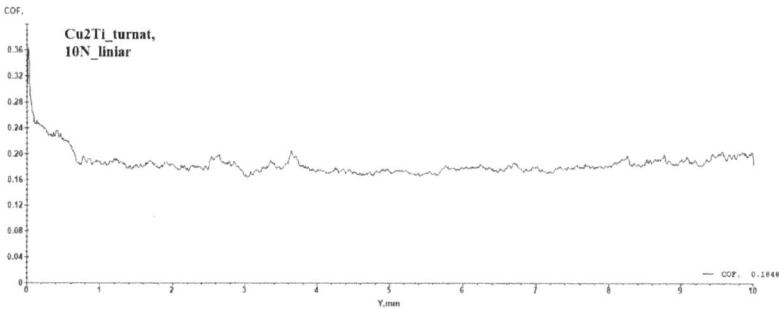

(b)

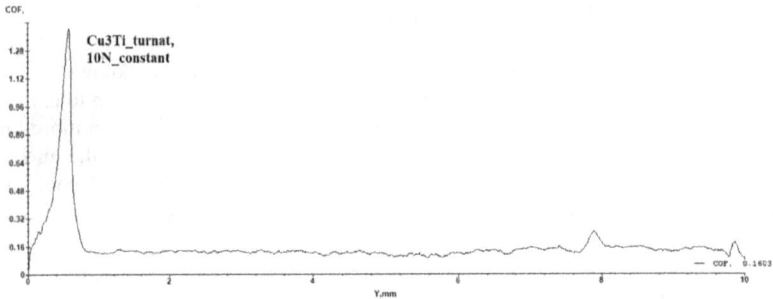

(c)

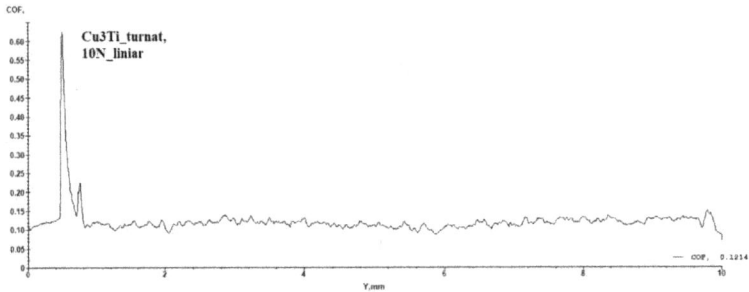

(d)

Figure 16 Coefficient of friction (COF) variation (a) and (b) for Cu2Ti and (c) and (d) for Cu3Ti.

Two types of stress were applied in this study. The values obtained for the coefficient of friction, acoustic emission, and friction force are shown in Table 5. The first type of stress was constant force, in which the support moved up and down to maintain a constant force. The second type of stress was linear force, in which a linear force (Fz) was applied as a function of time. The value of the force changed linearly from the initial value set by the user to the final value as a function of the duration of the experiment. This loading mode is often employed to determine the critical load.

Table 5. COF, AE and Ff parameter values for experimental samples.

Alloy/ feature		COF	AE	Ff
Cu2Ti_cast	10N_constant	0.77	0.008	7.7
	10N_ linear	0.18	0.002	1.01
Cu3Ti_cast	10N_constant	0.16	0.001	1.6
	10N_ linear	0.12	0.009	0.64

The highest coefficient of friction was identified in the Cu_2Ti cast alloy under constant stress conditions of 10 N. In the case of the Cu_3Ti alloy, an increase in stiffness was observed, resulting in a substantial decrease in both the coefficient of friction and the frictional force, regardless of the type of stress applied. With respect to sparking, materials exhibiting a lower coefficient of friction can be utilised; however, the formation of hot sparks is influenced by multiple factors.

4.3.2 *Severe wear analysis of Cu-Ti samples after explosion sparking tests*

CuTi alloys constituted of plates, clamped on a steel substrate, were subjected to a 16,000 cycle forced wear test as outlined in Chapter 3. The material degradation occurred within a potentially explosive enclosure between the CuTi alloys and a support with several carbon steel plates. Following the friction process (Fig. 17 (a)-(f)), the CuTi plates underwent various forms of wear, ranging from local oxidation of the alloy to removal of material both in the direction of wear and perpendicular to it. Some of the CuTi plates were plastically deformed as a result of the impact, while others were damaged close to complete destruction. During the wear process, no hot sparks were observed, thereby indicating that this alloy can be regarded as non-sparking for industrial applications in potentially explosive atmospheres.

The wear of the steel elements can be observed in Fig.s 17 (g) - (i), which show a transition of the CuTi alloy on the steel substrate through the aggressive wear that occurred, as well as the chipping of the steel in some areas following the tests.

(a)   (b)

Figure 17. Images of CuTi plate surfaces after severe wear test (16,000 contacts) to generate sparks that cause explosions (a) CuTi plates after wear on the fastening system, (b) the 6 rolled CuTi plates after severe wear (three each of Cu2Ti and Cu3Ti alloys), (c) different wear areas on the plates, (d) - (f) details of the different wear areas, (g) - (i) details of the steel plates used for Cu-Ti alloy wear.

The experiments were conducted under conditions of advanced wear, simulating all types of contact, from a fine contact at the beginning to an intense contact at the end. This approach covered most situations of alloy exploitation under experimental conditions. At the microscopic level (Fig. 18), the surface of CuTi plates was investigated by scanning electron microscopy. Material overlaps due to intense friction, macro and micro cracks, pores and oxidised zones were observed.

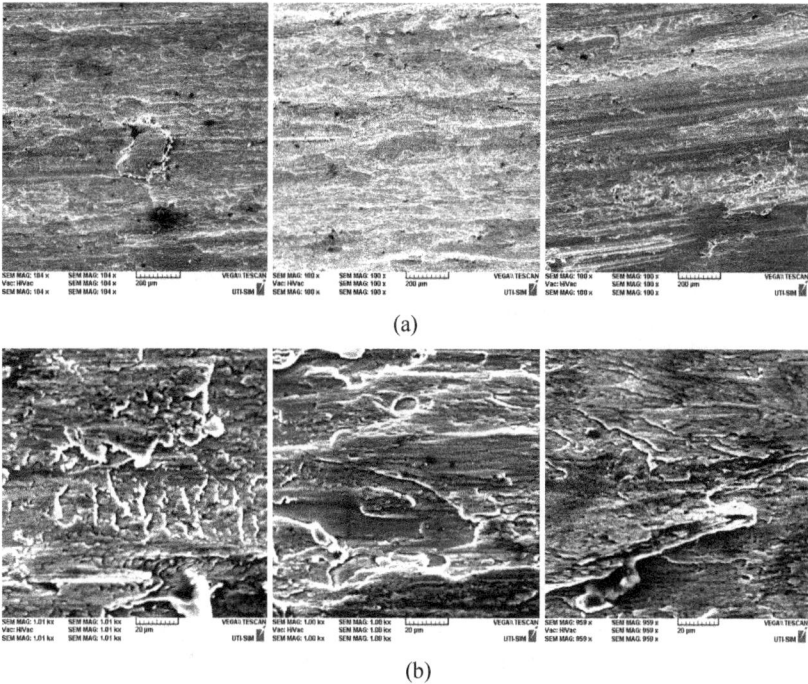

(a)

(b)

Figure 18. SEM images of CuTi plate surfaces after severe wear test for different areas (a) 100x and (b) 1000x.

As indicated in Fig. 19, an analysis of the chemical composition of the surface of CuTi plates after the wear tests revealed the presence of oxygen and carbon in addition to the fundamental elements of the alloy, i.e. copper and titanium.

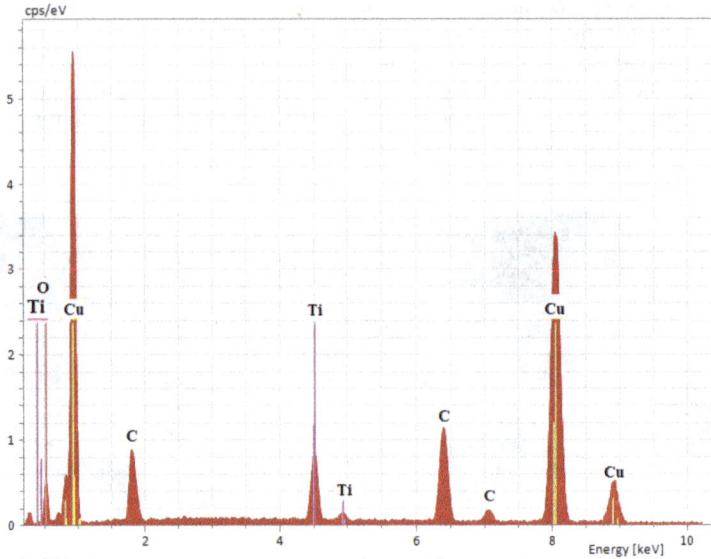

Figure 19. Energy spectrum identified by EDS analysis on the surface of CuTi plates after wear test and spark formation.

The chemical composition of the alloy indicates a slight oxidation of the surface, as observed in Table 6. The result of accentuated wear is a reduction in the percentage of copper on the surface, with the appearance of oxygen.

Table.6. Chemical composition of the surface of severely worn plates.

Area/Elements	Cu %		O %		Ti %	
	wt	at	wt	at	wt	at
Area 1	94	85	3.26	11.7	2.7	3.3
Area 2	91.5	77.2	5.95	19.9	2.53	2.83
EDS Error %	2.2		1.3		0.2	

The percentage of titanium remained consistent in comparison to the percentage of copper; this finding serves to substantiate the hypothesis that the titanium oxide formed on the surface exhibits superior resistance.

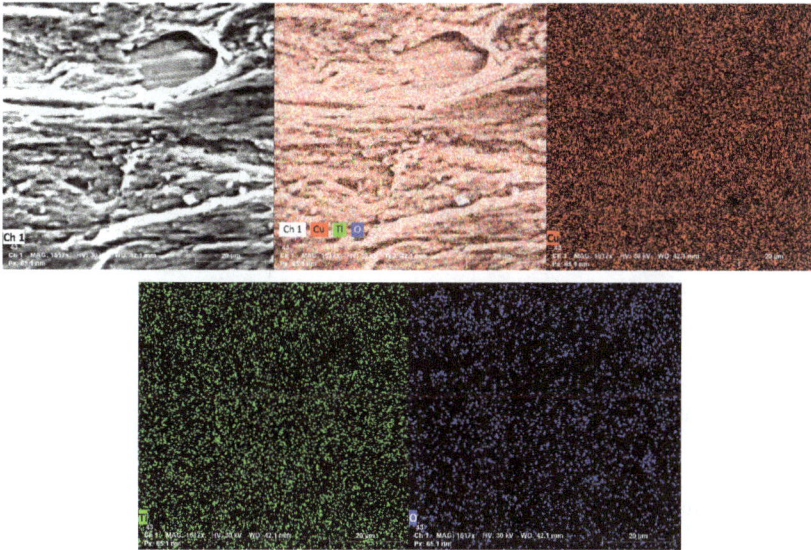

Figure 20. Main element distributions on the worn surface of CuTi alloy.

Fig. 20 presents the element distributions, from which a homogeneous surface wear with surface material breaks, oxides and pores is observed. The material breaks can be attributed to the direction of the stress forces, surface defects of CuTi alloys or stress steel, or accumulation at grain boundaries of intermetallic compounds.

An increase in the percentage of titanium from 2 to 3 wt% resulted in the formation of the intermetallic compound $TiCu_4$ in the alloy, in addition to the eutectic observed in Cu_2Ti. Although the formation of $TiCu_3$ was also identified by XRD, EDS tests did not provide confirmation. Its metastable state will lead to its transformation into the intermetallic compound $TiCu_4$. Titanium has been identified as solubilised in the copper matrix and in intermetallic compounds (mainly $TiCu_4$) positioned close to the grain boundaries.

Cu-Ti alloys have been identified as a viable solution for applications in potentially explosive environments. The process of rolling the alloy resulted in the hardening of the matrix. This was achieved by the formation of a greater number of grain boundaries and intermetallic compounds. These compounds were smaller in size when compared to their cast state, and were distributed homogeneously within the copper matrix. Furthermore, the titanium was solubilised.

Following a thorough analysis of the experimental results, several general conclusions can be drawn:

In order to ensure the safety of workers in special working conditions, it is necessary to establish specific working rules and to utilize appropriate materials. The materials selected, with their specific characteristics, have the capacity to make a significant contribution to improving working conditions, reducing the risk of explosion or securing the lives of workers in potentially explosive atmospheres;

Alloys in the CuAlBe and CuTi systems have been identified as a suitable solution for metallic elements proposed for use in potentially explosive atmospheres and can meet a wide range of requirements in this area;

Several experimental CuAlBe and CuTi-based CuAlBe and CuTi-based alloys have been produced by classical casting according to new compositions;

Both types of alloys can be processed by specific thermo-mechanical processes;

Both types of alloys can be hardened by alloying, heat treatment or thermo-mechanical processing;

The severe wear experiments conducted through 16 000 cycles and including material transfer from the tested alloys (CuAlBe and CuTi) did NOT produce hot sparks causing explosions, regardless of the tested alloy recipes;

CuAlBe alloys (approximately 2.1 GPa) have a higher hardness than Cu-Ti alloys (approximately 1.7 GPa). Regarding the Young's modulus (elasticity), Cu-Ti alloys have values several times higher than CuAlBe alloys;

Apart from the cast Cu2Ti alloy, all other CuAlBe and CuTi alloy compositions exhibited coefficients of friction (COF) close to each other (about ~ 0.2);

Both Cu-based alloys (CuAlBe and CuTi) showed similar behaviour in the saline electrolyte solutions in which they were tested (3.5% NaCl), forming a partially protective surface layer of reaction compounds consisting of oxides, chlorides and salts. While the corrosion current was substantially modified by mechanical rolling processing in CuAlBe-based alloys (a decrease of about 10 times compared to the cast alloy), this difference was much smaller in CuTi alloys. Cast CuTi alloys had higher resistance to electrochemical corrosion than cast CuAlBe alloys, but lower polarisation resistance than rolled CuAlBe alloys.

References

[1] I.S. Batra, A. Laik, G.B. Kale, G.K. Dey, U.D. Kulkarni, Microstructure and properties of a Cu-Ti-Co alloy, Mater. Sci. Eng. A 402 (2005) 118-125. https://doi.org/10.1016/j.msea.2005.04.015

[2] Y. Jiang, Z. Li, Z. Xiao, Y. Xing, Y. Zhang, M. Fang, Microstructure and properties of a Cu-Ni-Sn alloy treated by two-stage thermomechanical processing, JOM 71 (2019) 2734-2741. https://doi.org/10.1007/s11837-019-03606-5

[3] X. Wang, C. Chen, T. Guo, J. Zou, X. Yang, Microstructure and properties of ternary Cu-Ti-Sn alloy, J. Mater. Eng. Perform. 24 (2015) 2738-2743. https://doi.org/10.1007/s11665-015-1483-4

[4] M.A. Morris, M. Leboeuf, D.G. Morris, Recrystallization mechanisms in a Cu-Cr-Zr alloy with a bimodal distribution of particles, Mater. Sci. Eng. A 188 (1994) 255-

265. https://doi.org/10.1016/0921-5093(94)90380-8

[5] A. Schiavi, C. Origlia, A. Germak, A. Prato, G. Genta, Indentation modulus, indentation work and creep of metals and alloys at the macro-scale level, Materials 14 (2021) 2912. https://doi.org/10.3390/ma14112912

[6] S. Nagarjuna, D.S. Sarma, On the variation of lattice parameter of Cu solid solution with solute content in Cu-Ti alloys, Scripta Mater. 41 (1999) 359-363. https://doi.org/10.1016/S1359-6462(99)00187-6

[7] C. Li, X. Wang, B. Li, J. Shi, Ya. Liu, P. Xiao, Effect of cold rolling and aging treatment on the microstructure and properties of Cu-3Ti-2Mg alloy, J. Alloys Compd. 818 (2020) 152915. https://doi.org/10.1016/j.jallcom.2019.152915

[8] D. Hanoz, A.G. Settimi, L. Pezzato, M. Dabala, Effect of precipitation hardening on corrosion resistance of Cu-4.5Ti alloy, J. Mater. Eng. Perform. 30 (2021) 1306-1317. https://doi.org/10.1007/s11665-020-05353-0

[9] S. Ramesh, H.S. Nayaka, Investigation of tribological and corrosion behavior of Cu-Ti alloy processed by multiaxial cryoforging, J. Mater. Eng. Perform. 29 (2020) 3287-3296. https://doi.org/10.1007/s11665-020-04833-7

[10] S. Semboshi, T. Nishida, H. Numakura, T. Al-Kassab, R. Kirchheim, Effects of aging temperature on electrical conductivity and hardness of Cu-3Ti alloy, Metall. Mater. Trans. A 42 (2011) 2136-2143. https://doi.org/10.1007/s11661-011-0637-8

[11] P. Wei, L. Jie, Thermodynamics of Ti in Cu-Ti alloy investigated by the EMF method, Mater. Sci. Eng. A 269 (1999) 104-110. https://doi.org/10.1016/S0921-5093(99)00148-3

[12] S. Nagarjuna, M. Srinivas, High temperature tensile behaviour of a Cu-1.5Ti alloy, Mater. Sci. Eng. A 335 (2002) 89-93. https://doi.org/10.1016/S0921-5093(01)01945-1

[13] Y. Zheng, H. Zhao, S. Zhu, P. La, F. Zhan, M. Zhu, J. Sheng, H. Liu, Effect of rolling deformation on microstructure and properties of Cu-Ni-Mo alloy, Mod. Phys. Lett. B 35 (2021) 2150510. https://doi.org/10.1142/S0217984921505102

[14] K.V. Chuistov, Copper-titanium solid solutions as a new generation of high-strength age-hardening alloys, Usp. Fiz. Met. 6 (2005) 55-103. https://doi.org/10.15407/ufm.06.01.055

[15] W.A. Soffa, D.E. Laughlin, High-strength age hardening copper-titanium alloys: redivivus, Prog. Mater. Sci. 49 (2004) 347-366. https://doi.org/10.1016/S0079-6425(03)00029-X

[16] M. Ochoa, M.A. Rodríguez, S.B. Farina, Corrosion of high purity copper in solutions containing NaCl, Na_2SO_4 and $NaHCO_3$, Procedia Mater. Sci. 9 (2015) 460-468. https://doi.org/10.1016/j.mspro.2015.05.017

[17] L. Vrsalović, S. Gudić, D. Gracić, I. Smoljko, I. Ivanić, M. Kliškić, E.E. Oguzie, Corrosion protection of copper in sodium chloride solution using propolis, Int. J.

Electrochem. Sci. 13 (2018) 2102-2117. https://doi.org/10.20964/2018.02.71

[18] L. Prandtl, Proc. 1st Int. Cong. Appl. Mech., Delft (1924) 41-54.

[19] K. Koike, K.D. Clarke, A.J. Clarke, Microstructural evolution and mechanical properties of heavily cold-rolled and annealed Cu-3Ti alloys, JOM 71 (2019) 4789-4798. https://doi.org/10.1007/s11837-019-03838-5

[20] W.C. Oliver, G.M. Pharr, An improved technique for determining hardness and elastic modulus using indentation experiments, J. Mater. Res. 7 (1992) 1564-1583. https://doi.org/10.1557/JMR.1992.1564

[21] S. Bhaumik, V. Paleu, S. Datta, Tribological investigation of textured surfaces in starved lubrication conditions, Materials 15 (2022) 1-10. https://doi.org/10.3390/ma15238445

[22] M.F. Doerner, W.D. Nix, A method for interpreting data from depth-sensing indentation instruments, J. Mater. Res. 1 (1988) 601-609. https://doi.org/10.1557/JMR.1986.0601

[23] V. Krishnamoorthy, A.A. John, V. Paleu, Mapping acoustic frictional properties of epoxy-coated bearing steel using acoustic emissions, Technologies 12 (2024) 1-12. https://doi.org/10.3390/technologies12030030

[24] G. Pharr, Measurement of mechanical properties by ultra-low load indentation, Mater. Sci. Eng. A 253 (1998) 151-159. https://doi.org/10.1016/S0921-5093(98)00724-2

[25] C. Bejinariu, V. Paleu, N. Cimpoesu, Microstructural, corrosion resistance and tribological properties of Al$_2$O$_3$ coatings prepared by APS, Materials 15 (2022) 1-12. https://doi.org/10.3390/ma15249013

[26] X. Huang, G. Xie, X. Liu, H. Fu, L. Shao, Z. Hao, Influence of precipitation transformation on Young's modulus of a Cu-Be alloy, Mater. Sci. Eng. A 772 (2020) 138592. https://doi.org/10.1016/j.msea.2019.138592

[27] S. Nagarjuna, M. Srinivas, Grain refinement during high temperature tensile testing of Cu-Ti alloys, Mater. Sci. Eng. A 498 (2008) 468-474. https://doi.org/10.1016/j.msea.2008.08.029

About the Authors

Romeo-Gabriel CHELARIU

Ph.D. Eng.

gabriel_chelariu@yahoo.com

I am Ph.D. Romeo-Gabriel Chelariu, from "Gheorghe Asachi" Technical University of Iași, Faculty of Materials Science and Engineering. My doctoral research focuses on the development and characterization of non-sparking Cu-based alloys, such as CuAlBe and CuTi, designed for safe use in potentially explosive environments.

My scientific activity is mainly oriented toward the microstructural, mechanical, and surface analysis of advanced metallic materials obtained through casting and thermomechanical processing. I am particularly interested in the relationship between alloy composition, processing parameters, and performance under harsh operating conditions.

I have co-authored several papers published in international journals indexed in Web of Science and Scopus (*Materials, Archives of Metallurgy and Materials, Materials Today: Proceedings*), and have presented my research at national and international conferences such as BraMat, MSE, SESAM, and ICIR. Through my work, I aim to contribute to the development of innovative, safe, and efficient metallic systems for industrial applications.

Ramona CIMPOESU

I am associate professor Ph.D. Eng. at the "Gheorghe Asachi" Technical University of Iași, Faculty of Materials Science and Engineering. My research focuses on the corrosion behavior, mechanical performance, and surface modification of metallic and ceramic materials, with particular interest in biodegradable Zn-based systems and thermal barrier coatings obtained by plasma processing.

Throughout my academic career, I have authored and co-authored to over 70 ISI-indexed papers published in journals such as Applied Surface Science, Materials, Journal of Functional Biomaterials, Polymers, Crystals, and Metals, and I have participated in numerous international research projects related to advanced materials and coating technologies. I am also the author or co-author of eight specialized books published by recognized academic publishers.

My experience includes research stages at Université Lille 1 (France) and PROMES-CNRS laboratory (Font-Romeu, France) within the European project SFERA-III Horizon 2020, as well as teaching and collaboration activities under Erasmus+ programe.

I currently coordinate and participate in national and international research projects on biodegradable metallic materials, surface coatings, and plasma-assisted deposition techniques, aiming to improve materials performance in biomedical and industrial applications.

Gabriel-Dragos VASILESCU

Researcher I Habil.Ph.D. Eng. Gabriel VASILESCU

dragos.vasilescu@insemex.ro

I am researcher I at the National Institute for Research and Development for Mining Safety and Anti-Explosive Protection – INSEMEX Petroşani and Associate Professor of University of Petroşani, with 30 years of scientific researches and academic experience. I am head of laboratory "Explosive Materials and Pyrotechnic Articles" within the specialized department and a member of the Scientific Council of INSEMEX Petroşani INCD and I am doctoral supervisor since 2017, in the field of "Industrial Engineering" within the Doctoral School of University of Petroşani with nine completed doctoral thesis and seven ongoing doctoral students.

My main field of expertise is Industrial Engineering: I have published a total of 13 Books/Chapters and Monographs, over 150 research papers with more than 70 citations and a H index of 6-Web of Science / 6-Scopus / 10-ScholarGoogle. I worked at over 270 contracts of scientific research and development, in two of which I was project director/scientific responsible and seven of which project responsible. Furthermore, I own 4 invention patents in the field of Industrial Engineering, which were presented at numerous Invention Salons, winning medals.

I am a brand specialist in solving problems in the field of Health and Safety at Work, respectively: Modernization and accreditation of the technical infrastructure for the evaluation and certification of compliance in the field of industrial safety, including that specific to mining; The development and implementation of new concepts, methods and modern IT tools regarding the management of the professional/technological risk specific to industrial spaces with the risk of explosion and/or toxicity, including sites intended for specific operations with explosive materials such as civil explosives and pyrotechnic articles; The development of theoretical and practical solutions regarding the diagnosis/prognosis and appropriate management of technological/professional risks, as well as those of major mining accidents generated in industrial spaces with the risk of explosive and/or toxic atmospheres.

I am an active member of professional associations in the field, reviewer for scientific journals and in scientific committees at different congresses, as well as in the International Scheme of Trials for Interlaboratory Comparisons - RFCS 2015, (Royal Military Academy), Brussels, Belgium.

Costica BEJINARIU

Professor Ph.D. Eng. Costica BEJINARIU

costica.bejinariu@academic.tuiasi.ro; costica.bejinariu@gmail.com

I am full professor and researcher at the "Gheorghe Asachi" Technical University of Iasi, Romania, with over 35 years of academic experience. I was Vice Dean of Faculty of

Materials Research Foundations **186** (2026)   https://doi.org/10.21741/9781644903872

Materials Science and Engineering (2012-2024) and I am doctoral supervisor since 2009, with seven completed doctoral thesis and seven ongoing doctoral students.

My main field of expertise is Materials Engineering: I have published a total of 32 Books/Chapters and Monographs, over 300 research papers with more than 1800 citations and a H index of 21-Web of Science / 23-Scopus / 22-ScholarGoogle. I worked at over 50 contracts of scientific research and development, in five of which I was project director and two of which project responsible. Furthermore, I own 12 invention patents in the field of Materials Engineering, which were presented at numerous Invention Salons, winning medals.

I am an active member of professional associations in the field, reviewer for scientific journals and in scientific committees at different congresses. I am a member in the Committee of Materials Engineering, domain Engineering and Materials Science of the National Council for Attestation of University Titles, Diplomas and Certificates (CNATDCU), which is an independent consulting entity at a national level in Romania.

I am an Associate Member of the Academy of Romanian Scientists.